Ricardo Duran Baron
Jaime Luis Davila Torres
Luis Felipe Villero B

Obtaining essential oils and pectin from Valencia orange.

Ricardo Duran Baron
Jaime Luis Davila Torres
Luis Felipe Villero B

Obtaining essential oils and pectin from Valencia orange.

Orange grown in the municipality of Chimichagua Colombia

ScienciaScripts

Imprint

Cover image: www.ingimage.com

This book is a translation from the original published under ISBN 978-613-9-44126-6.

Publisher:
Sciencia Scripts
is a trademark of
Dodo Books Indian Ocean Ltd. and OmniScriptum S.R.L publishing group

120 High Road, East Finchley, London, N2 9ED, United Kingdom
Str. Armeneasca 28/1, office 1, Chisinau MD-2012, Republic of Moldova, Europe
Printed at: see last page
ISBN: 978-620-7-27424-6

Table of Contents :

Obtaining and evaluation of essential oils and pectin from by-products of the orange (*citrus sinensis osbeck*) Valencian variety grown in two co-regional zones of the municipality of Chimichagua to determine their application in the food industry.

Authors
RICARDO DURÁN BARÓN
JAIME LUIS DÁVILA TORRES
LUIS FELIPE VILLERO BERMUDEZ

DEDICATION

With special appreciation I dedicate this work to my dear old man, to my father José del Carmen Durán, who with his correct actions has taught me to live life in the best way, always keeping a positive attitude towards life and facing problems with great wisdom.

SUMMARY

Colombia does not produce pectin or the essential oils necessary to cover the demand generated mainly by the food and pharmaceutical sector, which has meant imports of pectin of 256,092 tons for the year 2006 and of essential oils of 1,794 tons for the period 20002003 (Agronet). This led to define the objective of the present work, which consisted of obtaining and evaluating the essential oils and pectin extracted from the orange (*citrus sinensis Osbeck)* Valencian variety cultivated in the department of Cesar to determine its application in the national food industry. Specifically, we worked with oranges produced in the municipality of Chimichagua in the villages of El Carmen and Higo Amarillo, whose municipality produces about 1050 ha of the 1554 ha existing in the department.

The pectin extraction process was developed using the acid hydrolysis method. The multiple range test was used to determine the significant differences between the means of the treatments at a confidence level of 95%, the experimental design was carried out in the statistical analysis program Statgraphics Centurion XV.II; 8 treatments were applied, involving 3 independent variables with two levels: type of acid (HNO3 and HCl), pH (2 and 3), and hydrolysis time (30 and 50 minutes); a standard temperature of 98° C or boiling temperature was used for all treatments. On the other hand, for the extraction of the essential oil, the method of steam entrainment and conventional microwave-assisted hydrodistillation was used, and for quantification, gas chromatography coupled to mass spectrometry.

The results showed pectin yields of 9,620 ± 0.764% (b.s.) for the Los deseos farm and 9,968 ± 1.114 %(b.s) for the Nueva Esperanza farm; the pectin obtained from the Nueva Esperanza farm resulted in a high degree of esterification ranging from 67.657 ± 0.870% to 71.227 ± 0.721% with an average of 69.442% and a high methoxyl percentage of 13.004% and an approximate purity of 39.613% as galacturonic acid, in addition to an acceptable degree of gelification of 152° SAG. The yield of essential oils obtained with conventional hydrodistillation assisted with microwaves at 600 WY 10 min was 0.11% for the Los Deseos farm and 0.71% for the Nueva Esperanza farm at maturity stage 2 in both cases; the vapor entrainment method did not yield favorable results.

These favorable results allow us to conclude that both pectin and essential oils have the quality and quantity characteristics to be used in the food industry.

INTRODUCTION

Valencian orange production in the Department of Cesar covers approximately 1769 ha, basically in the municipalities of Chimichagua, Valledupar, Astrea, Chiriguaná, Bosconia and Manaure (Cesar in figures, 2009-2010), scattered in farms of less than 10 hectares, totally primary, with little agronomic management, without any added value and sought after in the markets for its juiciness and sweetness. Harvesting occurs in June and at the end of the year, which leads to lower sales prices, increased post-harvest losses of up to 50% and environmental problems due to the deterioration and rotting of the product, which could be used for the production of juices, nectars, essential oils, pectins and other compounds (CCI, 2006).

The industrial phase of the citrus production chain includes products such as juices, concentrates, nectars, purees, pastes, pulps, jellies and jams (Espinal CF, 2005). By-products in the juice industry, consisting of peels (albedo and flavedo), seeds, membranes and juice vesicles, represent approximately 50% of the weight of the original whole fruit (Marín RF, 2007). These by-products can be used as nutrients in animal feed, marketed in the form of *pellets*, but their prices are not high enough to provide profitability to the utilization process (Wilkins MR, 2007), so the development of alternative products with higher added value would benefit citrus fruit processors.

Essential oils (EOs) and pectins are some of the products that can be obtained from citrus by-products (Kim WC, 2004). EOs are used in the production of liquors, perfumes, toiletries, as odor masking agents for paints and rubber, and as raw materials for the production of pharmaceuticals, among other applications (Sustainable BioTrade, 2003).

Pectins are used in the food industry as thickeners, gelling agents, emulsifiers and stabilizers (Mesbahi G, 2005), and in the pharmacological field as antimetastatic, immunostimulant and anti-ulcer agents (Liu Y, 2006). In addition, pectin, being a soluble fiber, decreases the low-density lipoprotein fractions in the blood, without modifying the levels of high-density lipoprotein, which is good for human health (Liu Y, 2006).

Orange samples were collected from one hectare in production at the Los Deseos and La Esperanza farms. The physicochemical characterization was carried out on the juice and peel at different stages of maturity. The extraction of EO was carried out using the steam dragging technique and by hydrodistillation with microwaves and its analysis by means of gas chromatography coupled to masses, to determine its majority components. Pectin was extracted by the conventional method (acid hydrolysis) and its characterization was determined by the degree of esterification, degree of methylation, galacturonic acid content, pH, free acidity, percentage of ash, percentage of moisture, time and speed of gelation, gelation potential and degree of gelation.

The objective of this study is to determine the content and composition of Essential Oils, and the quantity and quality of pectins present in the by-products of the orange fruit, grown in the department of Cesar, in order to determine the application of these compounds in the food industry.

Chapter 1

1. PROBLEM STATEMENT

The Department of Cesar has 1,000,949 ha for agricultural development of transitory, semi-permanent and permanent crops, of which 1,769 ha are oranges (URPA Statistics, 1993; Cesar in Figures, 2009-2010). The oranges produced are marketed fresh, without any added value or agro-industrialization, since there are precarious storage and marketing systems for the agricultural product. In addition to post-harvest losses, this means that it is impossible to generate employment and increase income. If this product were transformed into one of the many products known as orange juice, frozen orange concentrate, orange soda, fruit soda with orange as an ingredient, orange marmalade, orange jams, orange concentrate, animal concentrate, packing liquid with orange concentrate as an ingredient, frozen pulp, it would reduce or eliminate harvest and post-harvest losses and give added value that would increase income and make orange production and processing economically and environmentally sustainable.

Citrus fruits have been produced and consumed in Colombia for many years, in fresh form as such or in juice or in processed form as jellies, jams, juices, preserves, etc. The industrial utilization of citrus fruits leaves as waste material a large proportion of the total weight of the fruit (approx. 50%), very rich in pectins and essential oil, which is not used to obtain other products for use in food or industry.

There is no production of pectins in Colombia and around 256,092 tons per year (2006) are imported at prices that fluctuate in international markets between $15-30 dollars per kilogram (Quimerco, 2005), which means a value of more than seven million dollars that leave the country and indicates the rapid growth of the use of these products by the local industry, basically in the dairy and beverage sector.
The world market for essential oils is US$ 1.3 billion. There is great development and commercialization potential for citrus essential oils. Colombia imported citrus essential oils between 19982002, some US$6 million and US$14138 million in exports ("Biocomercio sostenible 2003").

In view of the above, the proposed research seeks to answer the following question: Will the Essential Oils and Pectin extracted from the by-products of the orange (*citrus sinensis*) Valencian variety cultivated in the Department of Cesar have the optimal characteristics for its use in the national food industry?

Chapter 2

2. JUSTIFICATION

It is estimated that in 20 years we will have a population of close to 9 billion and that we will need twice as much food production as we do today, which means increasing production, increasing agricultural frontiers or being more effective depending on chemical inputs, transgenesis or biotechnology. This difficulty increases with the current trend of using current agricultural soils for biofuel production.

The national citrus agroindustry has shown a significant, although very small, development in recent years, as it faces problems with the supply of raw material that does not meet its requirements in terms of quality, prices at certain times of the year, and location. In addition, there is little integration between industry and agriculture, there is no certified plant material, there is a lack of research and technology transfer (development of varieties and qualities) in the agricultural and agroindustrial phase, as well as pest and disease prevention.

According to the data provided by the national agricultural survey for 2009, a production of 465,015 tons of oranges was achieved in an area of 30,096 hectares of productive age, with a yield of 15.45 t/ha (ENA, 2009); of which the department of Cesar participated with 1,554.0 hectares harvested, with a production of 15,087.0 tons and a yield of 7.8 t/ha (Cesar in figures 2009-2010).

The Department of Cesar focuses its economy on agricultural production as a historical line item due to the different thermal floors, the richness of its soils, the hydrography and the customs of its inhabitants and in contemporary times, the other line item of great importance is mining, basically coal, whose results, apart from the export volumes and the royalties it generates, are displacing the active agricultural soils, changing the vocation of the community and generating high environmental impacts. These two lines of business have something in common and that is that they do not add value to the products they produce or extract, i.e. everything remains in the primary sector, i.e. coal is sold as it is extracted, practically from the pit to the export port and in the case of agricultural products from the harvest to local or national markets and some international markets.

In the DEVELOPMENT PLAN, a Cesar for all (2008-2011), actions are proposed to promote the planting of 4,000 hectares of staple foods such as bananas, tubers (cassava, yams, malanga), vegetables and fruits; expansion of the Incentive for Rural Capitalization of Cesar (ICRC); adequacy of 6,000 hectares with irrigation; quality programs in the implementation of good agricultural and manufacturing practices in food processing; an agro-productive model for agribusiness and marketing of Valencian oranges in the center of the country.A model for the agro-industry and marketing of Valencia oranges in the center of the Department of Cesar, which, if implemented, would allow for sufficient fruit production, and therefore a study is essential to determine the physicochemical characteristics and how to take advantage of the byproducts of Valencia oranges (pectins and essential oils). Since this is the most promising species in the department, if it has the desired characteristics for its agroindustrialization, the orange industry would become an important line in the economy of the department and the region (Departmental Development Plan 2008-2011).

The industrial utilization of the orange leaves as waste material a large proportion of the total weight of the fruit, very rich in pectins and essential oils, which is not used to obtain other products for use in food or industry.

Since the country should direct its efforts towards the utilization of its natural resources and especially towards obtaining those products that are easy to obtain and whose supply depends entirely on other countries, it was considered important to know the quality of the pectins and essential oils that could be obtained from the orange Citrus sinensis osbeck grown in the municipality of Chimichagua.

The development of this project would positively favor growers of citrus fruits such as oranges, since this fruit can have a high agroindustrial potential, and profits would increase to the extent that the fruit is given an added value, so that products such as juices, nectars, orange flavoring, jams, etc. are introduced into the market and the by-products or waste from agro-industrialization are used to obtain essential oils and pectins. Additionally, it could be used for the elaboration of perfumes, toiletries and raw material to produce pharmaceutical products, the results would be an input for the municipal and departmental administration for the promotion of a larger area of the crop. In addition, it contributes to the pectin and polysaccharides seedbed and to the agroindustrial optimization research group.

Chapter 3

3. THEORETICAL FRAMEWORK

3.1 . ESSENTIAL OILS

In the Far East, the history of essential oils began. The technological bases of the process were first conceived and employed in Egypt, Persia and India. However, it was in the West that the first scopes of its full development emerged. Experimental data on the methods employed in ancient times are scarce and vague. Apparently, only the essential oil of turpentine was obtained by the methods known today, although it is not known exactly what it was.

Until the Middle Ages, the art of hydrodistillation was used for the preparation of floral waters. When essential oils were obtained on the surface of the floral water, it was commonly discarded as an unwanted by-product.

The first authentic description of the hydrodistillation of essential oils was made by Arnold of Villanova (1235(?¡¡¡) - 1311), who introduced "the art of this process" in European therapy. Bombastus Paracelsus (1493 - 1411), who introduced "the art of this process" in European therapy.

1541) established the concept of the *Quinta Essentia*, that is, the most sublime extractable fraction technically possible to obtain from a plant and constitutes the drug required for pharmacological use.

There is evidence that the production and use of essential oils did not become general until the middle of the 16th century. The physicist Brunschwig (1450 - 1534) mentions only 4 essential oils known during that time: turpentine, juniper, rosemary and lavender (Günther, 1948).

With the advent of the steam engine and the use of steam boilers for manufacturing industries in the 19th century, hydrodistillation became a large-scale industrial process. Two types of generators emerged: live heat generators, where the boiler is part of the same vessel where the plant material is processed and works at atmospheric boiling temperature. And, steam boilers, which are not part of the equipment and work in a wide range of flows and temperatures for saturated steam. During the 20th century, efforts were made to improve the mechanical design of stills, minimize the high energy consumption required and adequately control the process (Günther, 1948; Al Di Cara, 1983; Heath and Reineccius, 1986).

Some recent research on obtaining essential oils from by-products of the extraction of mandarin, orange, grapefruit and grapefruit juices, using the technique of hydrodistillation assisted by microwave radiation, allowed the identification of its components by gas chromatography coupled to mass spectrometry, calculating their relative amounts and obtaining yields of essential oil of 0.23% (Jennifer P. 2009).

The essential oil of orange peel of the species cultivated in the region of Labateca, in Norte de Santander Colombia, *Citrus sinensis* valenciana variety, was obtained by hydrodistillation assisted by microwave radiation (HDMO) and analyzed by High Performance Gas Chromatography (HPLC), identifying as the major volatile component of the essential oil the oxygenated monoterpene limonene with 90.93%. (X. Yáñez Rueda, 2007).

A study conducted on the essential oil of the cashew orange, recovered from the peel of the fruit, which is used in the flavoring, cleaning agent, cosmetic and perfume industries, reports a higher oil yield by steam extraction, especially as the flow rate and pressure increase. In addition, gas chromatography identified the following as the main components of the extracts: benzaldehyde, terpinene, limonene, linalool, camphor, benzyl acetate, nerol, linalyl acetate and geranyl acetate (Grosse R, 2000).

3.2 PECTIN

Pectin was discovered in 1790 when Vauquelin found for the first time, a soluble substance from fruit juices. French scientist Braconnot continued Vauquelin's work and found that "a substance widely available from living plants and already observed in the past, had gelling properties when acid was added to its solution". He called it "acid pectin" from the Greek "pectos" meaning solid, coagulated (The Apple. The Pectin, Herbstreith).

There is production of pectin from valencia orange peel by acid hydrolysis and precipitation with ethyl alcohol. The pectin obtained presented a good appearance with gelling capacities comparable to the commercial standard. The yield obtained in the process was 10% on a dry basis for a pH close to 2.0 and a hydrolysis time between 30 and 40 minutes. To improve the yield, a second hydrolysis of the residue obtained once the initial mixture has been filtered can be carried out (Devia J., 2003).

A study carried out on passion fruit peel, comparing the conventional method with "steam entrainment", in which steam entrainment is carried out before acid hydrolysis with citric acid and hydrochloric acid, precipitation with alcohol and subsequent purification, showed a 15% increase in yield (15.63 on a wet basis) for the latter.

The variables that most influence the extraction process are: pH (1.5), extraction time (30 minutes) and type of reagent (citric acid with sodium hexametaphosphate). Pectin with a low degree of methoxylation was obtained, which forms a pre-gel in a range of 45-55° Brix, so it is used in the manufacture of soft jams, diet jams, nectars and ice creams, among others (Gaviria, N., López, L., 2005).

Microwave assisted extraction from dehydrated apple pulp was optimized using the response surface methodology, having as independent variables the extraction time, pH of the hydrochloric acid solution, liquid solid ratio and power, obtaining a yield of 0.157 grams of pectin per gram of dehydrated pulp with optimum values of 20.8 min, pH of 1.01, liquid solid ratio of 0.069 and power of 499.4 Watts (Wang S., 2005).

Using coffee processing residues (mucilage and pulp), pectin extraction was carried out using the conventional method at laboratory scale, obtaining yields of 4% (dry basis) and 10% (wet basis) of pulp and mucilage, respectively (Sierra L., Bolaños S., 2006).

Pectin was extracted from banana peel clone hartón by acid hydrolysis with hydrochloric acid in solutions of pH 2 and 3 for 60 minutes at a temperature of 85°C, obtaining pectin with a low degree of methylation and low gelation power. The yields on dry basis of pectin at pH 2.0 were 20.68% with a standard deviation of 3.32 and at pH 3.0 7.65% with a standard deviation of 1.41(D'addonisio R., 2008).

Comparing the extraction of pectins from valence orange, tangelo orange and grapefruit peels by conventional means in aqueous medium, with and without acid, the best yield was obtained with acid

(10.5% dry base), obtaining a pectin with a high methoxyl index (Gómez Z., Juan F., 1998).

3.38 THEORETICAL ASSES

3.3.1 RAW MATERIAL: *Citrus Sinensis Osbeck*

The raw material for the study is sweet orange peel, whose name is *Citrus sinensis osbeck, a* variety native to Asia. It belongs to the genus Citrus of the Rutaceae family, which comprises about 1600 different species; the citrus family being the most important, with approximately 20 species.

The fruits, which belong to the category of Hesperidia, are those that contain the fleshy matter between the endocarp and the seeds; they are widely used both for direct consumption and in industry, where they are used from food production to cosmetics.

Hodgson classifies oranges into four groups: sweet, navel, blood and non-acidic oranges (Morín, 1985).

Sweet oranges are the great majority of the commercial varieties existing worldwide, among them: Valencia, Pineapple, Hamlin, Parson Brown and native or criollas.

Taxonomy and Morphology:
Family: *Rutaceae*
Genus: *Citrus*
Species: *Citrus sinensis osbeck*
Growth habit: Small (3 - 5m); not very vigorous branches.
Flowers: Slightly aromatic; single or grouped with or without leaves. The buds with leaves are the ones that give the best fruits.
Fruits: Hesperidium (see figure 1) Consists of the exocarp (flavedo; presents vesicles containing essential oils), mesocarp (albedo, pompous and white) and endocarp (pulp, presents trichomes with juice). The Navel variety has supernumerary fruits (navel), which are small fruits that appear inside the main fruit due to a genetic aberration. Only 1% fruit set occurs, due to the natural exsorption of flowers, small fruits and closed buds.

To maintain a higher fruit set percentage, it is advisable to refresh the canopy by sprinkler irrigation, resulting in a slowing of growth, so that the fruit load is greater and smaller in size. The phenomenon of parthenocarpy is quite frequent (pollination is not necessary as a stimulus for fruit development). There are trials indicating that cross-pollination would increase fruit set, but the consumer does not want oranges with seeds. Some suffer from cell apomixis (an embryo is produced without fertilization).

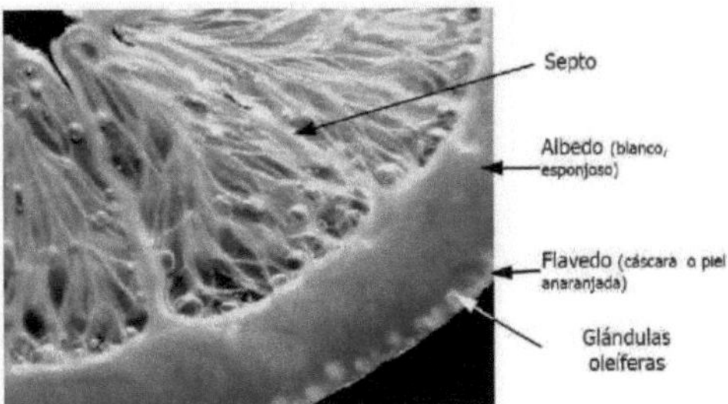

Figure 1. Structure of the orange

3.3.2 ORIGIN

The sweet orange tree comes from the southeastern regions of Asia, specifically from southeastern China and the Malay archipelago. It has been cultivated in southern China for thousands of years, from where it spread throughout Southeast Asia. The Arabs introduced the bitter orange tree in Europe through the south of Spain in the 10th century. On the other hand, the first known sweet oranges in Europe seem to have been introduced by the Portuguese from India at the beginning of the 16th century.

The orange is said in Latin "Aurantia", for its golden color, in Dravidian language (from India) "Narayan", which means "inner perfume". In Arabic, from the Persian language, "Narendj".

Citrus cultivation spread from Europe to the United States, where there are flourishing growing areas in Florida and California, to South America, where Brazil enjoys the highest share of the world market for oranges and orange juice, to South Africa and parts of Australia. Currently, the orange tree is one of the most widespread fruit trees worldwide, with the main producing countries being Brazil, the United States, Spain (Valencia, Murcia, Seville and Huelva), Italy, Mexico, India, Israel, Argentina and China.

1.2.1 COMPOSITION OF THE FRUIT

Table 1. Nutritional value of orange in100g of edible substance.

Element	Content	Element	Content
Water (g)	87.1	Citric acid (mg)	980
Protein (g)	1	Oxalic acid (mg)	24
Lipids (g)	0.2	Sodium (mg)	0.3
Carbohydrates (g)	12.2	Potassium (mg)	170
Calories (Kcal)	49	Calcium (mg)	41
Vitamin A (U.I.)	200	Magnesium (mg)	10
Vitamin B1 (mg)	0.1	Manganese (mg)	0.02
Vitamin B2 (mg)	0.03	Iron (mg)	0.4
Vitamin B6 (mg)	0.03	Copper (mg)	0.07
Nicotinic acid (mg)	0.2	Phosphorus (mg)	23
Pantothenic acid (mg)	0.2	Sulfur (mg)	8
Vitamin C (mg)	50	Chlorine (mg)	4

Source: http://www.infoagro.com/citricos/naranja.htm

Table 2. Physical and chemical composition of orange peel

Major components (%)	Dry matter	90
	Protein	6
	Carbohydrates	62,7
	Fats	3,4
	Fiber	13
	ashes	6,9
Minerals (%)	Calcium	2
	Magnesium	0,16
	Phosphorus	0,1
	Potassium	0,62
	Sulfur	0,06
Vitamins (mg/kg)	Hill	770
	Niacin	22
	Ac.	14,9
	Pantothenic	6
	Riboflavin	22,2
Amino acids (%)	Arginine	0,28
	Cystine	0,11
	Lysine	0,2
	Methionine	0,11
	Tryptophan	0,06

Demain and Solomon, 1986

1.3.4 ORANGE PRODUCTION

In the world

Oranges, with 63 million tons per year produced worldwide, are the most consumed fruit and represent the third most widely cultivated fruit after bananas and grapes. Today they are cultivated on all continents, provided they have a favorable climate, i.e. abundance of sun, water and low humidity.

The world's leading producer of oranges is Brazil, which uses most of its oranges for domestic consumption and orange juice production. Brazil is followed by the United States (Florida, California, Texas and Arizona) and China. Spain would be the main producing country in Europe and the first exporting country in the world. Other producing countries are, in order of importance, Mexico, which has greatly increased its production in recent years, Italy, India, Egypt, Israel, Morocco and Argentina (FAOSTAT, 2009).

Projections for oranges assume a slowdown in the expansion of orange production. The main reasons are severe disease problems in Brazil and Florida and reduced plantings in other parts of the Western Hemisphere due to the delayed effect of low prices in the past.

Expected orange production in 2010 is 66.4 million tons, about 14 percent higher than that obtained in the 1997-99 period. The projected annual growth rate at 1.12 percent is considerably lower than the 3.46 percent growth rate recorded between 1987-89 and 1997-99. The projected production should be used as fresh products (36.3 million tons) and as processed products (30.1 million tons), with the use of processed products projected to increase marginally.

Orange production in developed countries is projected to increase at an annual rate of 0.6 percent, with most of that growth coming from the United States. In Europe, there will be little change, with a small increase in Spain offset by declines in Italy and Greece. South Africa should continue to increase its production, taking advantage of supplying the northern hemisphere with off-season products. In Israel, production will continue to be affected by population growth, which will compete with citrus and produce crops for land and water use. Japan's orange industry will also continue to decline in the long term as the availability of imports increases.

Projections indicate that developing country production will increase at an annual rate of 1.23 percent. Over the next ten years, Brazil is likely to experience a significant contraction in production as the effects of disease and low producer prices are felt. By 2010, however, the Brazilian industry should recover, with production back to the levels recorded in the late 1990s, and maintain its dominance in the world market for processed oranges. Mexico is exposed to the citrus tristeza virus, which has already appeared in the Yucatan Peninsula. Mexican producers, mainly small growers, have not been able to take advantage of the preferential access to the U.S. market offered under NAFTA.

Smaller orange exporting countries in the Western Hemisphere, such as Argentina, Cuba, Belize and Costa Rica, should find market opportunities as major producing regions are involved in an adjustment process. Despite the trade embargo imposed by the United States, Cuba has expanded its orange production and processing capacity. The commercial sector in Belize and Costa Rica has also undergone a consolidation process that should lower costs.

Orange-producing countries in Asia are expected to continue to increase their production, which will be consumed almost entirely in domestic markets. Projections indicate that China will displace Mexico as the third most important orange producing country, and India will challenge Spain for fifth place. However, the sheer size of the domestic markets of these two countries suggests that virtually all production will be consumed domestically. Turkey is an exception, given its competitiveness in the European market, due to its geographical location and its association with the EC through a customs union. The Mediterranean countries of Morocco and Egypt should also benefit from their proximity to Europe.(FAO, 2012).

In Colombia

In 2009, Colombia had a total orange area of 27,343 hectares with a production of 223,919 tons and a yield of 9.93 t/ha (ENA, 2009).

In the department of Cesar

The department of Cesar for the year 2009 presented a total of 1,769 hs of orange planted with a total harvested of 1,554 hs, with a production of 15,087 Ton and a yield of 7.8 Ton/Has (Cesar in Figures 2009-2010).

1.3.5 PECTIN

Pectin is defined by Kertesz (1951) as water-soluble pectinic acids of varying degree of methylation, which are capable of forming gels with sugar and acid under certain conditions.

Pectin is a reversible colloid of lyophilic type, comprising a long chain of polygalacturonic acid molecules with carboxyl groups partially esterified with methyl alcohol, it is associated with sugars, hemicellulose, calcium and magnesium (Braverman, 1967). This compound belongs to the second group of polysaccharides, the heteropolysaccharides, which are located in the middle lamina between the cells and the primary cell wall of fruits and vegetables (Saldarriaga, 1974) and in combination with cellulose are responsible for the structural properties of fruits and vegetables.

During fruit ripening, pectin is broken down into sugars and acids, so the quantity and quality of pectin depends, among other things, on the age and ripening of the fruit. The softening of some fruits during ripening is partly due to pectinolytic enzymes: pectinmethyl esterase and polygalacturonase (Francis, 1975).

Their chemical composition, structurally they are mainly composed of galacturonic acid chains (see figure 2) linked in 1-4 bonds. The acid function is more or less esterified with methanol. The rhamnose molecules (methylpentose) are intercalated in the polygalacturonic chain by 1-2 and 1-4 bonds, producing an irregularity in the structure of the chain. It also has more or less long side branches (arabanes, galactans) linked to the secondary alcohol functions (see Figure 3) (Puerta, 2001).

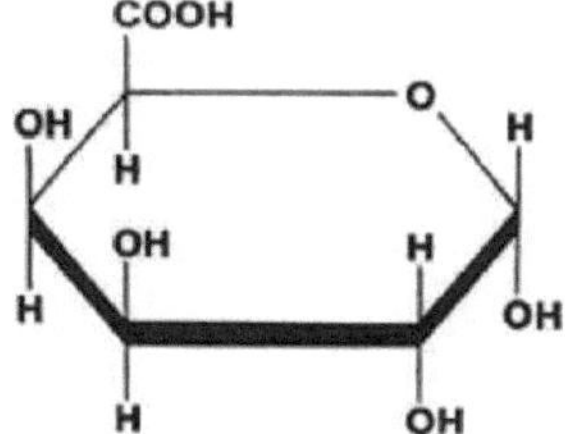

Figure 2. Structure of D - Galacturonic acid (Hercules, n.d.)

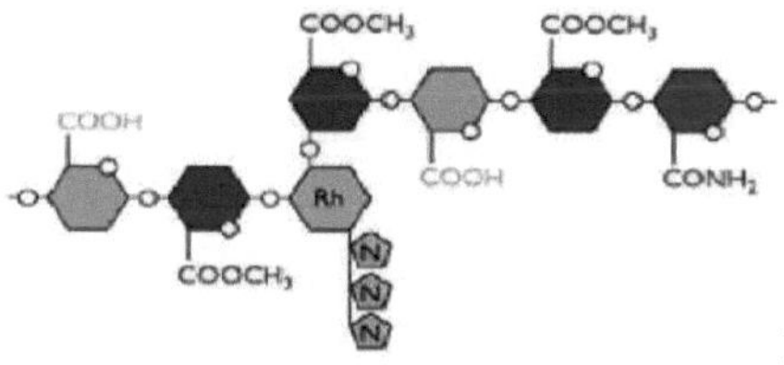

Figure **3.**

ture of pectin (Obipektin, 2005).

Properties of Pectins

The properties of pectins may vary according to the raw material used for their extraction:

Acid group: COOH Ester group: COOCHa Amide group: CONH2

Rh = Rhamnose
N = Neutral sugars (arabinose, galactose)

Gel-forming capacity or degree of gelation of pectins, which is defined as the number of grams of sugar with which one gram of pectin forms a gel of standard firmness, under controlled conditions of acidity and soluble solids. The grams of sugar required to form the gel is expressed as SAG degrees (Giraldo, 1991).

Pectin is the gel-forming agent, while solids (sugar) and acid are the modifying agents that achieve the physical transformation of pectin, converting the syrup into gel (Ramirez, 1980).

Good quality commercial pectins have gelling degrees between 150 and 300° SAG.

Its molecular weight is in a wide range between 2,500 to 1,000,000 gr/gr-mol depending on the source of extraction and the derivatives of the pectic substances found.

The degree of esterification is defined as the percentage of uronic carboxyl groups that are esterified with methanol. The determination of this percentage requires the measurement of the content of methoxyl ester and anhydrous uronic acid (Ramirez, 1980). It allows the determination of the gelation capacity of pectin.

Pectin solutions can have high or low viscosity values, depending on the quality and raw material used in extraction (Ramírez, 1980). Those whose viscosity is higher are more used for the elaboration of jams. Pectin solutions generally show lower viscosities compared to other gums and thickeners.

Different concentrations of sugars or calcium, as well as pH, affect viscosity in different ways.

Purified and dried pectin is soluble in hot water (70-80°C) up to 2 to 3%, forming viscous lumps on the outside and dry on the inside, which is why it is always mixed with sugar, buffer salts or other chemical substances (Ramírez, 1980). It must be fully dissolved to avoid heterogeneous gel formation. The commercial pectin shows a great affinity with water, which practically leads to the creation of clots at the time of dissolution, the addition of sugar reduces the solubility preventing the appearance of these. (Hercules, sf).

Acidity. Pectin solutions are stable under acidic conditions (between pH 3.2-45) even at high temperatures, also for a few hours at room temperature under more alkaline conditions, but degrade rapidly at high temperatures.

Stability in solution. Most of the reactions to which pectin is subjected tend to degrade it. In general, maximum stability is obtained at pH 4. The presence of sugar in solution has a certain protective effect; at high temperatures and low pH values the degree of degradation increases due to hydrolysis of glycosidic bonds; de-esterification is also favored at low pH. For high methoxyl pectins, when the temperature or pH increases, what is called p-elimination begins, which is a chain break and a rapid loss of viscosity and gelling properties. Low methoxyl pectins show better stability under these conditions (Hercules, sf).

Reactions with other electrically charged hydrocolloids. Pectin reacts with positively charged macromolecules, e.g. proteins at pH lower than its isoelectric pH (Hercules, sf).

3.3.5.1 CLASSIFICATION

According to the chemical changes and transformations due to fruit ripening.

- **Protopectin:** Discovered by Fremy in 1840, it is the native form of pectin. It is a water-insoluble polymer found in the early stages of formation and maturation of plant tissues. It consists of partially methylated sugars, in particular galacturonic anhydrous units linked together. It is contained in an unknown form in plant tissues (Giraldo, 1991).

- **Pectin**: When pectic substances become soluble they are known as pectin. As the fruit ripens, Protopectin is converted into pectin and pectinic acids by the action of an enzyme called pectinmethylesterase, which solubilizes it (Saldarriaga, 1974).

- **Pectinic acids:** It is a colloidal polygalacturonic acid with a methoxyl content of less than 4% that forms gels with sugars and acids. They form salts such as sodium pectinate (Eskin, 1971).

- **Pectic acid**: A high molecular weight polymer with galacturonic acid units, it contains no methoxyl groups, so all carboxyl groups are free (Eskin, 1971).

- **Pectinates:** Pectin salts (Giraldo, 1991).

- **Pectates:** Salts of pectinic acid (Giraldo, 1991).

According to the degree of esterification (methoxylation)
Commercial pectin is currently classified according to the degree of esterification (DE), which is defined as the percentage of carboxyl groups esterified with methanol (number of moles of methanol per hundred moles of galacturonic acid) (Yates, 1999).

Accordingly, pectins can be classified into three groups according to their degree of esterification.

- **High methoxyl (HM) pectin:** It has in its molecule some galacturonic acid units esterified above 50% and is presented as the methyl ester of galacturonic acid (Figure 4). This pectin is soluble in water, since it has almost all the carboxylic groups esterified with methanol (methoxylated), for this reason it is called high methoxyl pectin (Ortiz, 2002).

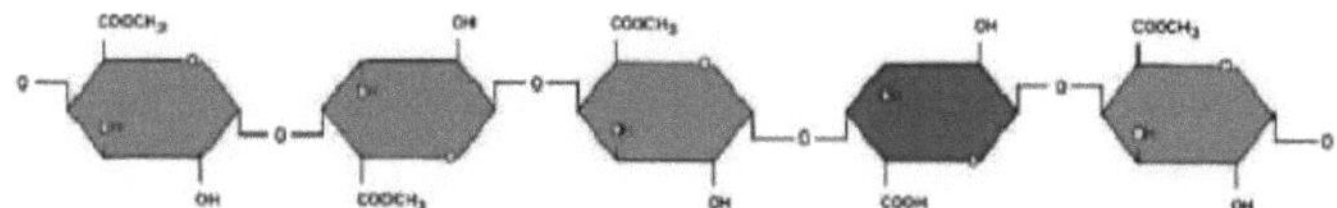

Figure 4. High Methoxyl Pectin (Calvo, 2006)

Pectins with a high degree of esterification are classified as follows:

a) **Fast gelling:** They have a degree of methoxylation of at least 70%. The strength of the gels formed depends on the molecular weight and is not influenced by the degree of methoxylation; the higher the molecular weight, the greater the gel strength (Braverman, 1974).

b) **Slow gelling:** It has a methoxylation degree of 50 to 70%, the amount of acid required is almost proportional to the number of free carboxyls (Braverman, 1974).

- **Conventional low methoxyl (LMC) pectins:** Low methoxyl pectins can be obtained from high methoxyl pectins by a chemical de-esterification process. These pectins have a degree of methoxylation below 50% (see Figure 5) and form a gel in the presence of calcium ions and other polyvalent cations. The amount of pectin required for the formation of these gels decreases with the degree of methoxylation. The strength of ion-bound gels depends on their degree of methoxylation and is greatly affected by the molecular weight of the pectins (Braverman, 1974).

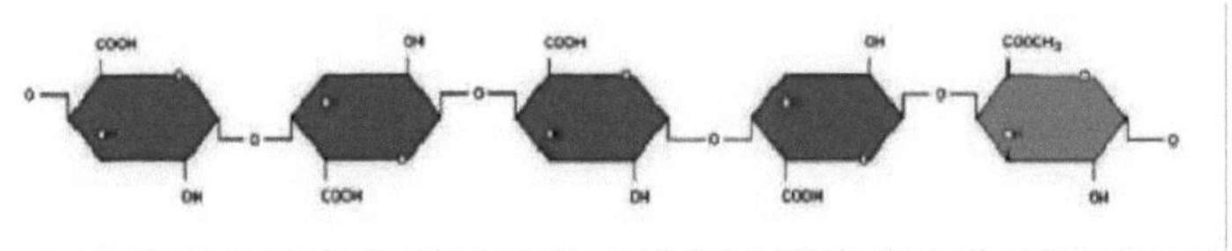

Figure 5. Low methoxyl pectin (Calvo, 2006).

Low esterification pectin is classified according to its reactivity with calcium ions:

a) **Fast Pectin: It** has a high reactivity with calcium ions, an esterification degree of approximately 30% and an amidated group content of 20%.

b) **Medium fast pectin:** It has a medium reactivity with calcium ions, contains an esterification degree of approximately 32% and an amidation degree of 18%.

c) **Slow Pectin: It** has a medium reactivity with calcium ions, contains a degree of esterification of approximately 35% and a degree of amidation of 15%.

- Low methoxyl amidated pectins: Similarly, a type of pectin called amidated pectin can be obtained from HM pectins (see Figure 6). This transformation is possible by treating HM pectin with ammonia under alkaline conditions in alcoholic suspensions, in this process the ester is replaced by the amide group (Yates, 1999).

Amidated pectin has a better ability to form gels in the presence of calcium compared to regular low methoxyl pectin.

They can have over 25% of amidated groups in their structure and this changes the temperature and texture characteristics. In this type of pectin some of the remaining galacturonic acid groups have been transformed into amide. Useful properties may vary with the ratio of ester to amide units and with the degree of polymerization (Hercules, food gums division, sf.).

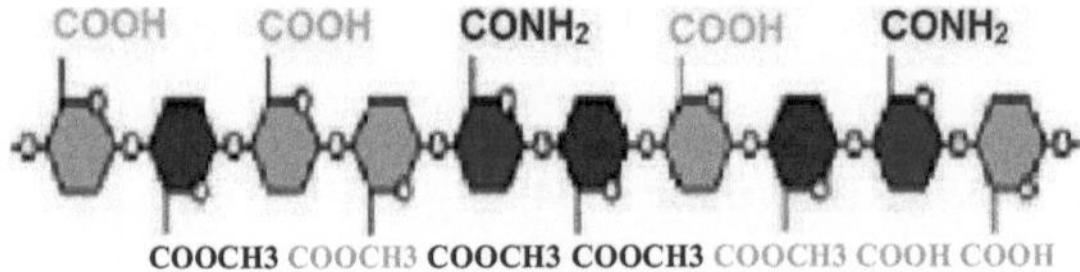

Figure 6. Low-esterification amidated pectin (Obipektin, 2005).

According to the chemical composition of the polymeric chain
They are classified into Galacturonans or polymers of galacturonic acid (Miranda, 1993); Ramogalacturans or mixed polymers of galacturonic acid and rhamnose (Miranda, 1993); Arabinogalactans or mixed polymers of arabinose and galactose (Miranda, 1993) and Arabinans or polymers of arabinose (Miranda, 1993).

3.3.5.2 EXTRACTION METHOD

For the conventional extraction method, the following operations are carried out

The selection of the material should take into account the quality of the plant material to be used, i.e. material without fungi, without rotten parts and the level of maturity of the plant, since unripe material tends to contain a higher percentage of pectins (Devia, 2003).

In the inactivation of pectic enzymes, it should be taken into account that pectinolytic enzymes are extracellular metabolites produced by some microorganisms (mainly yeasts and fungi) such as *Bacillus pumilus* 29 (Cabeza, 2003). These enzymes are mainly used in the juice production industry; those that use pectic compounds as natural substrates are called pectinolytic enzymes (Yegrs, 2001).

In order to make the extraction process more efficient, it is necessary to inhibit the enzymatic activity of the pectic enzymes, in addition to eliminating dirt and/or microorganisms present in the selected material. Inactivation is carried out by boiling a mixture of ground plant material and water (Devia J, 2003).

The process of pectin extraction is done through an acid hydrolysis, for which there are two methods. The open method consists of heating the solution of plant material (ground), some acid and water under agitation and in a vessel open to the atmosphere. The closed method consists of heating the same solution in a vessel with a condenser attached (Devia, 2003).

Heating time is a critical variable in the extraction process as is temperature (Pagan, 1998) and the acids normally used in this extraction process are sulfuric acid, citric acid, nitric acid or hydrochloric acid (Devia, 2003).

Once the hydrolysis process is completed, the material is filtered through a sieve to separate the solid and the liquid solution (Devia, 2003).

Next comes the precipitation process, which consists of separating the pectin from the acid solution, for which salts or alcohols are used (preferably the latter, since pectins are used in the food industry). In this precipitation it is recommended to use a volume of alcohol equivalent to 80% of the solution to be precipitated (Devia, 2003).

Finally, drying should be carried out at a temperature of approximately 40^0 C for 12 hours or in the open air for several days (Devia, 2003).

3.3.5.3 PECTIN CHARACTERIZATION
The physicochemical characteristics of isolated pectins are dependent on the extraction processes, and although pectins are composed primarily of unbranched chains of 1,4-D-galacturonic acid, there

are no simple chemical derivatives that appear to characterize pectic substances (Joslyn, 1962; McCready, 1970). Useful information is obtained by well-established and commonly used arbitrary analyses and these are primarily physical (Genu., 1979). The most important are the anhydro-galacturonic acid (AAG) content, ash percentage, degree of esterification and some data on the molecular weight of the pectin which may be reflected by viscosity and gelling power or degree of jelly (Ullman, 1950; McCready, 1970).

Chemical determinations

Pectic substances differ in their chemical properties from other polysaccharides mainly due to the presence of large amounts of free carboxylic groups or groups esterified by methyl groups (Pilnik and Voragen, 1970).

- **Purity as anhydrogalacturonic acid (AAG) or pectic substance content**

Pectin purity is indicated by its AAG and ash content. The former indicates the percentage of other organic materials present, generally neutral polysaccharides, and the ash, inorganic impurities (Rouse and Knoor, 1970).

The determination of pectic content in foods presents difficulties due to the structure of the pectin molecule (Ibarz, 2006). The vegetable matrix usually contains large amounts of starch, sugars, cellulose and other carbohydrates associated with pectin that interfere with its analytical determination (Kitner and Van Buren, 1982). This represents a problem in routine analyses where fast and accurate techniques are required (Ibarz, 2006).

The methodology to quantify pectins in plant materials involves a previous extraction and purification stage and the final determination is carried out with different techniques, being colorimetric methods the most used due to their greater selectivity with respect to volumetric and gravimetric methods, and their greater simplicity in relation to chromatographic and electrophoretic techniques (Carbonell et al., 1990).

Several colorimetric methods are known for the quantification of pectic substances, among them the carbazole method (McCready and mCComb, 1952) and the m-hydroxyphenylphenol method (Blumenkrantz and Asboe-Hansen, 1973). These techniques are based on the reaction of 5-formyl-2furanocarboxylic acid, from the hot action of sulfuric acid on galacturonic acid, with a colorimetric reagent to obtain a colored product that absorbs at a certain wavelength maximum (Ros et al., 1992). Barazarte et al. (2007) found that both methods reproduce anhydrogalacturonic acid values quite well when comparing experimental measurements with the corresponding standards.

The carbazole method has been the most widely used as a routine analysis (Visciglio and San Juan, 2000). However, recent works prefer the m-hydroxyphenylphenol method for its greater sensitivity and specificity in the presence of sugars and phenolic components (Ibartz et al., 2006). Regardless of the technique applied, the highly complex and heterogeneous structure of pectins requires selectivity and accuracy of analytical quantification methods (Carbonell et al., 1989; Rodríguez et al., 1992; Willats et al., 2006).

- **Degree of esterification (methoxylation)**

McCready (1970) defines the degree of esterification (% esterification) as the percentage of carboxylic-uronide groups that are esterified with methanol, over the total uronide contained in the pectin. On the other hand, the degree of methoxylation (% methoxylation) is the ratio of methoxylated to total galacturonic acid groups in the understanding that galacturonic acid is only partially esterified (Genu P. Co., 1979).

The degree of esterification or methoxylation is an important parameter in estimating the behavior of a pectin in terms of its dispersion rate in aqueous solutions, gelation time of the gel produced with it, sensitivity to polyvalent cations and its ability to form normal or low-solids gels (Rouse et. al., 1964: Pilnik, 1970);
McCready, 1970; Francis, 1975). Most of the pectins in commercial use are high methoxyl, and their content varies from 7% to 12% MeO (42.1 to 73% in basea16.3% MeO).
Determination of these functional groups can be achieved by basic hydrolysis of the functional groups followed by titration of the remaining alkali. The methoxyl content is equivalent to the alkali consumed in the hydrolysis (McCready, 1970).

- **Equivalent weight**

Equivalent weight values in pectic substances are used to calculate the % AAG and % esterification (Owens et al., 1952). Olsen (1939) noted that the equivalent or combination weight decreases progressively with the demethoxylation of pectin, as does its rate of gelation. Thus, equivalent weight is a comparative parameter for determining the quality of pectin from a given source (McCready, 1970; Rouse et. al. 1970). The determination of the equivalent weight can be achieved by direct titration of pectins with standardized alkali (Owens, 1952).

- **Ashes**

The ashes indicate the inorganic impurities of the pectins which are probably combined with the carboxylic groups to form salts of pectinic acids. A small part of the pectin is occluded in the precipitate and is not combined (Rouse and Knoor, 1970).

Determination of physical properties.

Although pectins of equal degrees of esterification can be produced by any of the methods described, and the products appear chemically identical, the physical properties are often very different.
The establishment of standardized methods for the physical characterization of pectin has been an arduous and complex task. This becomes apparent when observing the diversity of products in which a given pectin from the same production batch can be applied. It was not until after 20 years of research that the California Fruit Products Exchange Laboratories did research for physical analysis and adopted the "jelly grade" as the single most important standardizing factor for commercial pectins (Joseph and Baier, 1949; IFT, 1959).

- **Determination of the gelling power of pectin.**

As mentioned, pectin chain length is directly related to molecular weight and this can be determined relatively by measuring the increase in viscosity of a pectin solution. This is a useful index when comparing pectins of similar methoxyl content, especially in production control. However, there is a viscosity limit beyond which there is no increase in gelling properties, so the final test, for any pectin, should always be the jelly test (Francis and Bell, 1975).
Pectins have a jelly grade according to their agility to form gels. This has been defined as the sugar holding power consisting of the parts by weight of sugar that will hold one part by weight of pectin, under standard conditions forming a gel of satisfactory properties (IFT. 1959). The higher the jelly grade the less pectin is required for standard gel formation, but this is not the only criterion for the selection of a given pectin.
The standardization of gel preparation methods and the design of a simple and reliable test to measure their firmness was a difficult task and several alternatives were presented. The problem to be solved was the establishment of a method that would control the most important factors in the formation and firmness of pectin gels and that would be representative and useful for the various applications that a given pectin might have (Cox and Hibby, 1944; Joseph and Baier, 1949).
In 1949, the research department of the "California Fruit Growers Exchange" carried out a thorough investigation of the methods of preparation of pectin gels known to date. They established what is known today as the "standard gel" and its preparation method (IFT, 1959).
Simultaneously, it was necessary to submit for consideration a test that would qualify the gel

produced. There are two accepted possibilities for testing the jelly:

1- The gel may be subjected to conditions that exceed its elastic limit, where it is broken or permanently deformed.
2- The test may involve the observation of a permanent, non-measurable change during the test period.

Irreversible deformation or gel rupture is influenced by several factors that are difficult to evaluate separately. Moreover, these factors are highly dependent on gel production and handling techniques (Joseph and Cox, 1947).

The apparatus proposed by Locwood and Hayes (1931), known as the "Exchange stiffness tester", was developed to measure gel stiffness by observing and measuring, after a given time, the reduction in height due to expansion occurring in a standard test gel of a given height. The apparatus was chosen as the best test instrument by the Institute of Food Technologist (1959), although some members accepted it with reservation as the method has limitations in giving information on certain important gel characteristics, but in general the apparatus used was accepted as a reproducible method of gel preparation, giving a reproducible index of the ability of pectin to retain sugar by forming a gel. The method fails to give information on the optimum pH range for a given pectin nor does it describe the surface tension of the gel nor does it evaluate the usefulness of the pectin in certain types of gels requiring different preparation conditions than the standard, as well as other points of importance for food products (IFT, 1959).

- Time and temperature of gelation.

The gelation or settling time is the period immediately after a batch of jelly has been mixed until it settles into a firm gel. This time is a function of the temperature of the medium, the cooling rate of the gel, the presence of polyvalent ions, the pH, the gel solids, the amount of pectin used and other factors that modify the formation and firmness of the gel (Francis, 1975; Flores, 1966). According to its purity, degree of esterification and degree of polymerization, pectin will tend to be fast, medium or slow gelling.

Generally, high molecular weight pectins with esterification degrees of 65% or more will be fast settling and those with degrees between 50 and 60% will be slow settling (McCready, 1970). At lower degrees of esterification, the pectin is low methoxyl and will tend to be fast-settling (Owens et al., 1952; Pilnik, 1970). Degradation by aging of high methoxyl pectin will result in the progressive tendency to low methoxyl behavior. The presence of polyvalent ions drastically decreases the settling velocity of low methoxyl pectin.

In the manufacture of preserves, very fast gelling pectins are required, so that the solid particles are embedded throughout the gel and do not float to the surface of the container (Genu P. Co. 1979). Slow gelling pectins may take an hour to gel and the gel will form at temperatures below 70°C: these are used in gels that have to be packaged before they begin to coagulate. The gelling temperature increases with all factors that increase the gel firmness. Certain buffer salts such as citrates or sodium tatrate act to retard the rate of gelation. Pectin manufacturers must adapt their product to the requirements of the food producer and this is achieved by mixing pectins of various types and grades with addition of buffer salts if necessary (Francis, 1975).

The gelling time is commercially known as the solidification time at 30°C. For its determination, the gel preparation follows the same rules as for the determination of firmness and jelly grade. The time is started by filling a beaker of standard dimensions with the liquid gel at 96°C. The beaker is immersed to the rim in a water bath at 30°Cc and rotated with gentle turns at timed intervals. When the gel at the bottom of the beaker begins to gel, it is seen to rotate in the opposite direction to that of the printed rotation. The test time ends when the phenomenon is observed on the surface (Joseph and Baier, 1949; IFT., 1959; McCready, 1970). Using this method, slow-settling pectins gel in 3.25 to 4 minutes, while fast-settling pectins gel in less than 10 seconds. Food manufacturers are familiar with the interpretation of the results of this test, since the rate of gelling will vary greatly as some factors vary (see Table 3) (Joseph and Baier, 1949).

Gelling time in minutes

Air (20 25°c)	Water bath (30oc)
1.0	0.90
2.0	1.65
3.0	2.25
4.0	2.80
5.0	3.25
7.0	4.00
10.0	5.10
14.0	6.30

Joseph and Baier, 1949

- Dehydrated pectin moisture.

Pectin prepared by any process should be dehydrated and stored in cold, airtight conditions. Pectin absorbs moisture rapidly up to 15% of its sin depending on the humidity of the environment. Moisture should be determined when the pectin reaches equilibrium moisture (McCready, 1970).

3.3.5.4 PECTIN APPLICATIONS

In the food industry

The use of pectin in high sugar content jams is one of the best known applications to one of the largest markets for pectin.

High methoxyl pectins, in decreasing order of their esterification percentage and speed of gel formation, are commercially classified into the following types (Navarro and Navarro, 1985).

The measurement of the gelling capacity of pectin sugar gels is standardized and measured in SAG degrees.

a) Ultra fast set 150° SAG (URS 150)°
b) Rapid set 150o SAG (RS 1500)

c) Medium rapid set 150° SAG (MRS 150°)
d) Slow set 150° SAG (SS 150°)

Their main characteristics are shown in Table 4 (Navarro and Navarro, 1985).

Table 4. Characteristics of high methoxyl (HM) food pectins.

Limit values

Feature	1RS 150	RS150° RS150	MRS 150° MRS 150	SS 150
% of steiification	74-77	71-74	66-70	58-65
gel formation (minutes)	1-3	4-8	15-25	30-120
Optimum gelling pH	3.1-3.4	3,0-3,3	2,8-3.1	2,6-2,9
pH of solution at 1% Total ash (%) Drying loss (%) Arsenic (ppm) Lead (ppm) Copper (ppm) Pathogenic germs Total germs per gram Appearance Grammometry	2,9-3,5	2,9-3. 5 approx. 12 less than 3 less than 10 less than 60 absence less than 1000 fine stag-colored powder rejection less than 1% on sieve opening = 0.31m	52.8-3,5	2,8-3,5

Navarro and Navarro, 1985

Within the EU (European Union) there are two standards, jam and extra jam containing with some exceptions a minimum of 30% or 45% fruit pulp respectively.

High quality jam also tends to be made with higher quality fruit, which implies less pectin. Table 5 (May, 1990) indicates which fruits require high, moderate or small pectin addition.

Table 5. Relative additions of HM pectin to jams and marmalades.

HIGH 2 0.3% HIGH 2 0.3% HIGH 2 0.3% HIGH 2 0.3%	**AVERAGE (approx. 0.3%)**	**LOW s 0.3% LOW s 0.3%**
Cherry	Apricot	Apple
Pear	Blackberry	Black currant
Peach Pineapple	North American raspberry	Red currant Guava
Raspberry Strawberry		Plum

Jams are often made from depectinized fruit concentrates and will then require a large amount of added pectin from the type of fruit in question.

High methoxyl pectins are used only in standard jellies above 60% soluble solids. Some countries now allow sugar reduction to 30-55% soluble solids or even lower.

The selection of the correct pectin is important (the one that is effective at lower soluble solids content and more sensitive to calcium will be the pectin that should be used) but fruit content is also important. Sometimes, especially at low soluble solids content, it is necessary to add calcium salt to obtain the best result. Occasionally neutral gums are added to reduce syneresis. pH control is very important.

A range of icings for pastry and custard can be prepared using a formulation with amidated low methoxyl pectin and calcium sequestrants such as diphosphates.

A growing area of fruit production in recent years has been the production of fruit bases for addition to yogurts and similar products. These fruit bases containing 20-60% sugar have been made with modified starches as thickeners. In spite of their economical price, they present problems such as aroma masking and irregular texture and have been advantageously substituted by low methoxyl amidated pectin.

Pectin also has other uses in the dairy industries. High methoxyl pectin preserves dairy products from casein aggregation when heated to pH values below 4.3. This effect is used to stabilize drunk and UHT-treated yogurts and also for milk and fruit juice blends. It also stabilizes acidified milk drinks with soy and wheat-based products, where it prevents protein precipitation. Yogurt can be thickened by the addition of very low levels of low methoxyl amidated pectin.

Low-calorie beverages are very clear (texturally) and have the characteristic lack of mouthfeel provided by sugar in conventional soft drinks. Pectin can be used to improve the texture of such products and, thus, replace fruit pulp in such products.
In sorbets, ice creams and ice lollies, pectin can be used to control crystal size. In ice lollies it retains aromas and colors, which normally tend to come out of the ice structure.

Gelatin has been the traditional base for jelly desserts. They are formulated with low methoxyl amidated pectins that provide the right texture and freezing point.

Pectin forms are generally recognized as safe by the U.S. Food and Drug Administration (Food and Nutrition Encyclopedia, 1983) and their legal specifications are stated internationally in the food codex (Copenhagen Pectin).

In general terms, in the food sector, pectin has applications in the production of:
- Jams, jellies, preserves, flans, jellies and fruit compotes. Approximately 75% of the world production of pectin is used for this purpose (Ramírez, 1980).
- Slimming agents
- Canned fruits

- Synthetic powders for the homemade manufacture of gelatines.
- Preparation of natural juices, to reinforce the stability and viscosity of these products.

- Stabilizer in ice cream production.
- Manufacture of sauces.
- Stabilizer and slimming agent in salad dressings and milk drinks.
- Preservation of refrigerated and frozen foods, avoiding water leakage during thawing.
- Production of candies and confectionery
- Processing of the outer layer of sausages.
- Bread making, 5% pectin is used in relation to the weight of the flour. - Ice cream, desserts and milk industry.

It is used in medicine and pharmaceutical industry for tasks such as: Intestinal cleansing, Wound treatment, through the preparation of dressings and bandages, facilitating wound healing (Saldarriaga, 1974), Blood transfusions, as a substitute for blood plasma. In the preparation of insulin. In the preparation of penicillin, to reduce the degree of absorption. It prolongs the action of adrenaline, natural hormones, streptomycin, ephedrine, etc. Preparation of ointment for skin ulcers, treatment of traumatic shock, regulator of intestinal tract. For intravenous treatment of shock. Good blood agglutinant, for which it is used in the treatment of intestinal hemorrhages; however, it is not convenient to use the pectin solution in excess because it forms in the organism a compound of unknown nature (Ramirez, 1980).

In the cosmetics and toiletries industry, it is used in the preparation of toothpaste and as an absorbent in soaps.

In the metalworking industry it is used for hardening steel or other alloys, a solution of 0.2 to 4% can be used, pectin is more advantageous than oils because the operation is easily regulated by changing concentrations (Saldarriaga. 1974). It is also used in the coating of aluminum sheets.

In addition to the above, it is used in the plastics industry in the manufacture of foaming products as fining agents and binders. In the preparation of fibers. In the manufacture of cellophane paper and decorative ribbons and in the preparation of adhesive substances as a substitute for dextrin.

3.3.5.6 **PECTIN MARKET**

Worldwide producers include: Hercules Incorporated, CpKelco (Denmark), Danisco (USA), Herbstreith & Fox (Germany), Obipektin (Switzerland) and Degussa (Germany) (International Pectin Producers Association).

The amount of pectin consumed domestically is imported in its entirety because there is no pectin-producing company in Colombia, which implies high sales prices for this product.

In the market, pectin is offered in kilograms and according to its classification is its selling price. Low methoxyl pectin is offered at $60,000 plus VAT per kilogram, the "Rapid set" and "Slow set" have a selling price of $35,000 plus VAT per kilogram (data supplied by Bell Chem Internacional S.A., Medellín). The price is strongly influenced by the degree of soluble solids required for its gelling, the lower the amount of soluble solids required, the higher the price of pectin.

In the Colombian market there are different distributors of pectin at the national level, among which are the company Quimerco S.A. and Danisco Colombia Ltda.

Quimerco S.A. is a company that represents eleven leading foreign companies; it stores and distributes food ingredients, chemical raw materials and plastics for the adhesives, paints and food industries; its activity is concentrated solely in Colombia.

Quimereo S.A.'s pectin supplier is CPKelco, which has more than 2,000 customers in 100 countries and production facilities in North America, Europe, Asia and Latin America (Brazil).

Another important pectin producer with 25% of the world market is Danisco, which has production plants located in Brazil, Mexico and the Czech Republic.

Both Quimerco S.A. and Danisco Colombia Ltda. are headquartered in Bogotá, from where they distribute their products throughout the country. At the regional level, pectin is distributed by Bell Chem Internacional S.A. and Distribuidora Córdoba.

Pectin imports come from different countries such as Mexico, Argentina, France, Brazil, China, Denmark, Switzerland, Germany, Belgium, among others (Legiscomex, 2007).

Finally, it is important to highlight the total amount of pectin imported during these three years (2004-2006) as shown below (see Figure 7):

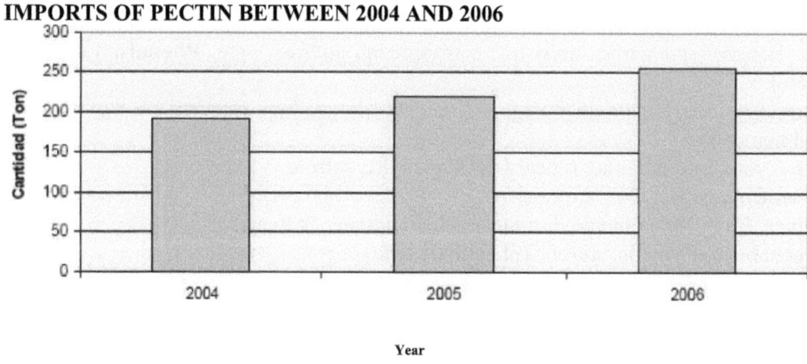

Figure 7: Total amount of pectin imported by Colombia during the years 2004 to 2006.

In 2004, 193,039 tons of pectin were imported, while in 2005, 27,197 tons more were imported than in 2004 and in 2006, 35,856 tons more were imported than in the previous year.

The pectin market in Colombia is basically focused on the food industry for the production of jams, candies, snacks, desserts and ice cream, among others. Specifically, low-esterification pectin is used in the production of soft spreadable jams, diet jams, yogurt, nectars, juices, sauces, and ice cream.

The consumer companies buy pectin according to the parameters of fast, medium fast or slow gelling, but most of them do not know whether they are high or low esterification pectin, and the advice given by the distributors about the product that best suits their needs plays a very important role.

3.3.6 ESSENTIAL OILS

Essential oils can be obtained from orange peel, which is one of the basic ingredients in the perfumery, food, agronomic and pharmaceutical industries.

Essential oils are formed in the green (chlorophyll-bearing) parts of the vegetable and as the plant grows they are transported to other tissues, in particular to the flowering buds. The exact function of an essential oil in a vegetable is unknown; it may be to attract insects for pollination or to repel harmful insects, or it may simply be an intermediate metabolic product.

Essential oils are volatile liquids, mostly insoluble in water, but readily soluble in alcohol, ether vegetable and mineral oils. They are usually not oily to the touch. Alicyclic and aromatic hydrocarbons, as well as their oxygenated derivatives, e.g. alcohols, aldehydes, ketones, esters, sulfur and nitrogenous substances, can be found in an essential oil. The most common compounds are biologically derived from mevalonic acid and are categorized as terpenes, the most abundant being monoterpenes (C10) and sesquiterpenes (C15). They are basic ingredients in the perfume industry and are used in soaps, disinfectants and similar products.
Compounds dissolved in essential oils can be classified as follows:
Esters. Mainly benzoic, acetic, salicylic and cinnamic acids. **Alcohols.** Linalool, geraniol, citronellol, terpinol, menthol, borneol.
Aldehydes. Citral, citronellal, benzaldehyde, cinnamaldehyde, cuminic aldehyde, vanillin.
Acids. Benzoic, cinnamic, myristic, isovaleric all in free state. **Phenols.** Eugenol, thymol, carvacrol.
Ketones. Carvone, menthone, pulegone, irone, fenchone, thujone, camphor, methylnonyl ketone, methyl heptenone.
Esters. Cynol, internal ether (eucalyptol), anethole, safrole.
Lactose. Coumarin.
Terpenes. Camphene, pinene, limonene, phellandrene, cedrene.
Hydrocarbons. Cymene, styrene (phenylethylene).

Citrus essential oils are insoluble in water, but become more soluble when used in low concentrations using alcohol as a solvent. They sometimes form dark solutions that clarify with difficulty. Hence, it is desirable to remove terpenes and sesquiterpenes. Two methods can be applied for this purpose, by fractional distillation at reduced pressure, or the extraction of the more soluble oxygenated compounds (main carriers of the odor), with diluted alcohol and other solvents.

In some oils there is a large amount of terpenes. This is especially true of lemon and orange oils, which contain up to 90% d - limonene in their normal composition. Not only are terpenes and sesquiterpenes of little value to the strength and character of the oils, but they also oxidize and polymerize rapidly at rest to form compounds with a strong, turpentine-like flavor. Furthermore, terpenes are insoluble with the low strength of the alcohol used as a solvent, so they form dark solutions that clear with difficulty. It is therefore desirable to remove terpenes and sesquiterpenes from many oils. This type of oil, for example orange oil, is 40 times stronger than the original and produces a clear solution in the diluted alcohol. The oil now has a slight tendency to become rancid, although it does not have the original freshness. Because each oil has a different composition, deterpenation requires a special process. Two methods can be applied, either the removal of terpenes, sesquiterpenes and kerosenes by fractional distillation at reduced pressure, or the extraction of the more soluble oxygenated compounds, with diluted alcohol or other solvents.

The main advantage of essential oils is that they can be used in any food and the FDA has considered them as GRAS (Generally Regarded as Safe) substances, as long as their effect is achieved with minimal change in the organoleptic properties of the food (Viuda-Martos *etal.*, 2008; Burt, 2004).

3.3.6.1 Physical and organoleptic properties of essential oils. Generally, essential oils are liquids at room temperature. Their volatility or capacity to evaporate on contact with air, at this temperature, differentiates them from fixed oils. Within aromatic compounds, the molecular weight is restricted to a maximum of 250g/mol so that the substances can volatilize (Pauli, 2001). They are easily alterable or sensitive to oxidation, although they do not become rancid like lipids. They have a tendency to polymerize, giving rise to the formation of resinous products, especially those containing unsaturated terpenic alcohols, varying their odor, color and viscosity. They are fatty oils, easily soluble in organic solvents, such as petroleum ether, chloroform, benzol or absolute alcohol; and almost insoluble in water, to which they communicate their odor (Pérez, 2006). The density of essential oils varies from 0.84 to 1.18 g/cm^3 ; most of them are less dense than water. They have a high refractive index, with an average of 1.5 and generally present optical activity (Perez, 2006).

3.3.6.2 Composition of essential oils

In the same species, the qualitative and quantitative composition of the essential oil can vary according to the part considered, the conditions of the environment in which they have developed, the time of the year, their particular genetic endowment, etc. (Fisher and Phillips, 2008). It can be said that more than 60 individual compounds can be obtained, but the major components constitute more than 85% of the essential oil (Coronel, 2004). Among the main compounds are terpenes, aromatic compounds derived from phenylpropane and various other compounds. The latter are found in small quantities and are organic acids such as acetic, valeric, isovaleric; coumarins and ketones of low molecular weight, etc. Perez (2006),

Essential oils from citrus fruits contain 85 to 99% volatile components that are a mixture of monoterpenes such as limonene, sesquiterpenes and their oxygenated derivatives, including aldehydes such as citral, ketones, acids, alcohols such as linalool, and esters (Fisher and Phillips, 2008).

In orange essential oil, 200 different chemical compounds have been found, of which 100 have been identified. Monoterpenes comprise up to 97% of the composition and the rest are alcohols, aldehydes and esters (1.8 to 2.2%). The main component of these oils is limonene, whose structure is shown in Figure 8. This compound can comprise up to 98% of the composition of orange essential oil (Fisher and Phillips, 2008).

CH_3

H_3C CH_2

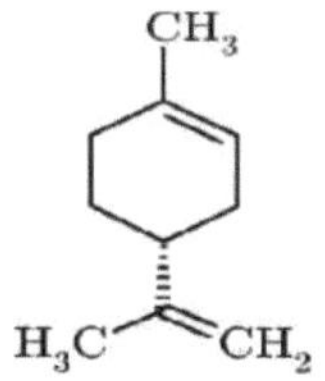

Figure 8. Chemical Structure of Limonene
(Merck Chemicals, 2009)

Limonene is a hydrocarbon classified within the cyclic terpenes, there are three forms: d, l and dl.

- It is a natural biodegradable solvent present in citrus fruits, with interesting chemical properties, pleasant aroma and qualified as safe and environmentally friendly.
- It can be used in pure form, mixed with other solvents or oils, or emulsified to make water-soluble cleaning products.
- It is insoluble in water (Stashenko, 2000).

Table 6 shows the technical characteristics for technical grade and food grade limonene.

Table 6. Technical characteristics of limonene

Feature	Technical grade	Food grade
Appearance	Yellow to white oily water liquid	Oily white water
Odor	Strong orange aroma	Mild orange aroma
Specific gravity(25°C)	0.838-0.843 g/cm^3	0.838-0.843 g/cm^3
refractive index(20°C)	1,4710-1,4740	1,4710-1,4740
Flashpoint (flash point)	46°C	45°C
Boiling point	154°C	163°C
Solubility in water	Insoluble	Insoluble
Vapor pressure(20°C)	2 mmHg	2 mmHg

3.3.6.4 Essential oil extraction

Essential oils are obtained through methods such as pressing, extraction, fermentation or distillation, achieving extraction yields ranging from 0.1 to 2%, with some exceptions, such as Chinese badian with a yield of 5% or clove, with more than 15% of essential oil (Costa-Batllori, 2003).

- **Steam distillation steam distillation extraction**

Steam distillation is a technique used to separate water-insoluble and slightly volatile substances from other non-volatile products mixed with them. It is the most commonly used, both industrially and in the laboratory, for the production of essential oils, since these have volatile compounds that can be carried away by water vapor; in addition, it is a simple and economical technique that is highly versatile when applied to different plant materials (Jiménez *et al.*, 2006; Ortuño, 2006).

The basis behind this extraction technique is given by the breaking of the plant tissue by the effect of the steam temperature (100°C) thus releasing the essential oil after a certain time (Sanchez, 2006).

Vapor entrained substances are immiscible in water, have low vapor pressure and high boiling point. When distilling a mixture of two immiscible liquids, its boiling point will be the temperature at which the sum of the vapor pressures of each liquid is equal to the atmospheric pressure (Jiménez *etal.*, 2006).

In this technique, the sample is placed in a vessel through which water vapor generated in another vessel is passed. A stream of water vapor is passed through the sample containing the compound to be extracted, which causes the sample to heat up and the vapor carries away the volatile components into a cooling system typical of a simple distillation. The distillate is collected, separated and purified (Perez, 2006).

- **Extraction by hydrodistillation with microwave-assisted Clevenger device.**

Microwave-assisted extraction (MAE) is a process that uses microwave energy to heat solvents in contact with a sample in order to partition analytes from the sample matrix into the solvent. It is very useful because it reduces extraction times, since the energy of the microwave or super-frequency magnetic field (2450 MHz) is mainly converted into heat in substances made of polar molecules and as a result of intensive vapor formation in the pore-capillary structure of the plant material large pressures develop altering the physical properties and on the other hand less deterioration of the plant tissue compared to conventional heating.

- **Other methods**

In addition to steam distillation and microwave-assisted hydrodistillation with Clevenger device, there are other methods for obtaining essential oils, such as cold pressing, hot fat extraction (enfleurage) and extraction with solvents, either petroleum derivatives, supercritical fluids or non-petroleum solvents (Ortuño, 2006).

3..3.7. METHODS OF ANALYSIS - EXTRACTION AND ISOLATION).

Isolation is performed using one or more chromatographic methods such as column chromatography, thin layer chromatography and HPLC. For column and thin-layer chromatography, silica gel is widely used as the stationary phase. As mobile phase pure or mixed apolar solvents are used such as: toluene-ethyl acetate 93:7, benzene, chloroform, dichloromethane, benzene-ethyl acetate 9:1, benzene-ethyl acetate 95:5, chloroform-benzene 75:25, chloroform-ethanol-acetic acid 94:5:1(Wagner, H., 1984), chloroform-benzene 1:1(Harborne, 1973; Stahl, 1969).

However, more efficient and faster separation techniques such as HPLC high performance liquid chromatography and gas chromatography (GC) are currently used (Dugo, 1992; Griffiths, 1992; Ingham, 1993), as well as "ON-LINE" HPLC-CG-MS combinations (Mondello, 1994; Mondello, 1996). These same methods are used for the analysis of flower essences (Barkman, 1997).

This last technique, thanks to the recent development of high resolution capillary columns, allows the analysis of complex mixtures present in essential oils, and the identification of the components from the retention times through the so-called Kovats Retention Indexes (Ik) (Denayer, 1994). These values are characteristic for each component and there are databases with the indices of many essential oil components.
Ik values are determined on two chromatographic columns, one polar (e.g. CARBOWAX 20M) and one non-polar (e.g. OV-101 also called DB-1).

Figure 9 shows the gas chromatogram obtained for orange peel essential oil (Blanco Tirado, 1995).

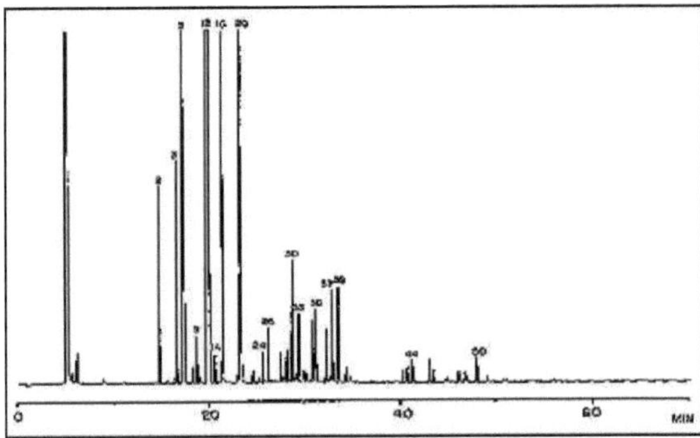

Gas chromatogram obtained for orange peel essential oil.

Additionally, the coupled technique Gas Chromatography-Mass Spectrometry, allows obtaining the mass spectrum of each component with which the molecular weight and structural information is obtained (Maat, 1992). There are also databases with the mass spectra of many components, so that the Kovats index (determined in two columns of different polarity) and the mass spectrum are criteria for the chemical assignment of many components of essential oils, not only monoterpenes but also other types of substances characteristic of these oils.

More recently, chiral chromatographic columns have been developed for the separation of optically active components (Ochocka, 1997), and methods have been developed for the combined HPLC-Mass Spectrometry and HPLC-NMR analysis of sesquiterpene mixtures (Vogler, 1998).

3.3.7.1. GAS CHROMATOGRAPHY (GC)

The gas chromatograph separates a mixture into its individual components which then enter one by one into the ion source of the mass spectrometer, or any other detection device.

The successful application of GC to the study of EOs has been possible due to the development of capillary columns that allow the separation of mixtures of different polarity or isomerism (monoterpenes, sesquiterpenes, etc.). The most commonly used stationary phases by GC for the analysis of essential oils are: CARBOWAX 20M [poly (ethylene glycol), polar phase] and OV-101 (DB-1, HP-1, HP-1, SP-30) [poly (dimethylsiloxane)], apolar phase (Jennings, 1980).

Retention time is not a completely reliable parameter for qualitative analysis, since it depends on many experimental variables, such as: type of column, entrained gas flow, amount of sample injected, etc. Kovats (Kováts, 1958) proposed a system of retention rates that serves as a basis for the qualitative analysis of the components of complex mixtures (tentative identification).

The Kovats retention indices system is based on a comparison between the position of the peak of an analyte in the chromatogram and the peaks corresponding to linear hydrocarbons, one of which elutes before the component of interest and the other after. The following mathematical relationship is used to calculate them, in the case of a chromatographic run with temperature ramp:

$I_K = 100*n + 100*[(T_x - T_1)/(T_2 - T_1)]$ Equation 1
Where:
I_K: Kovats index
n: Number of carbons of the standard hydrocarbon that elutes before the analyte of interest.
T_X : retention time of the analyte of interest.
T_1: retention time of the standard hydrocarbon eluting before the analyte of interest.
T_2: retention time of the standard hydrocarbon eluting after the analyte of interest.

3.3.7.2. MASS SPECTROMETRY (MS)

Mass spectrometry is a fast and sensitive method of analysis that allows the maximum amount of structural information to be obtained with the minimum amount of sample (10-6-10 -14 g).

The mass spectrum is a plot relating the masses of specific ions (more precisely, ion mass-to-charge ratio values, *m/z)* to their respective concentrations in the total ionic current (TIC) produced by ionization and fragmentation of the analyte molecules in the ionization chamber.

The mass spectrum provides information on the molecular mass, elemental composition of a substance (when using high-resolution MS) and, in some cases, allows the spatial structure of the molecule to be established.

The mass spectra of terpenes, the main constituents of essential oils, are in most cases very similar. Their identification by the MS method is mainly based on quantitative differences (ion abundances). However, it has been possible to find experimental criteria (ratio of peak intensities of molecular and characteristic ions, MS of the activated collision, kinetic energy released during metastable transitions, etc.) for the identification of terpenes (basically monoterpenoids) through their mass spectra (Konig, 1998; Adams, 1995).
In several works (Konig, 1998), the identification of sesquiterpenes is carried out based on their Kovat indices. The structure of a sesquiterpene is considered established only if its mass spectra and Kovats indices for two stationary phases (polar and apolar) coincide with those of the standard substance and with those of the databases of mass spectra (fragmentation patterns) and Kovats indices of these compounds (Jennings, 1980; Adams, 1995).

3.3.7.3. GAS CHROMATOGRAPHY COUPLED TO MASS SPECTROMETRY (GC-MS).

The coupling of a gas chromatograph with a mass spectrometer allows rapid and reliable analysis of complex mixtures, simultaneously determining how many and which constituents are present in the mixture and in what proportions. This method is highly sensitive and can detect components in picogram quantities.

The modern GC-MS system consists of three main blocks: gas chromatograph, mass spectrometer and data system. The computerized data system performs the conversion of the signals coming from the mass spectrometer into chromatograms and normalized mass spectra, allowing to obtain the structural information of each of the components of the mixture (Stashenko, 1998; Adams, 1995; Schreier, 1984).

3.3.7.4. APPLICATIONS OF ESSENTIAL OILS

Essential oils have medicinal properties and have been used since ancient times to cure diseases.

Modern science processes them to obtain specific drugs or remedies to prevent or cure various conditions in humans and animals. Based on their properties, they are widely used for the digestive, respiratory, nervous and circulatory systems. They are currently widely used in the world in aroma therapy.

These oils provide the food industry with characteristic flavors and aromas, widely used in bakeries, jams, confectionery, candies, soft drinks, ice cream, preservatives, cookies, dairy products, etc.
They are used in the chemical industry to provide aromas to cleaning products, such as air fresheners, perfumes, soaps, detergents, dishwashers, hospital products, as well as insecticides and disinfectants.
They are used in the cosmetics industry for the manufacture of colognes, perfumes, toilet soaps, various types of creams, shampoos, deodorants, hair conditioners and fixatives, etc.

They have diverse uses in agroindustry as natural insecticides, balanced feed for fattening pigs and poultry, tobacco flavoring, etc. Depending on the product processed, the residual plant material, once the oil has been extracted, is used as a feed supplement for cattle, pigs or poultry or as a natural fertilizer (Martínez, 2003).

Chapter 4

4 OBJECTIVES

4.1 General

To obtain and evaluate essential oils and pectin from by-products of the orange (citrus sinensis) Valencian variety cultivated in two co-regional zones of the municipality of Chimichagua to determine their use in the food industry.

4.2Specific

- Perform physicochemical analysis of Valencian orange fruit.
- Extraction of essential oils by steam stripping of the peel, membranes and remaining juice vesicles.
- Perform pectin extraction by the conventional method from the peel, membranes and remaining juice vesicles.
- To evaluate the essential oils and extracted pectin.
- Establish the application of Essential Oils and pectin in the food industry according to the results of the evaluation.

Chapter 5

5 METHODOLOGY

5.1 Bibliographic review

Databases, degree projects, books, articles, web pages, and technical visits were reviewed to provide the information required for this research.

5.2 Collection of plant material

Oranges of the species *Citrus Sinensis osbeck*, Valenciana variety, Rutaceae family, grown in plots located in the rural area of the municipality of Chimichagua, 250 km from Valledupar, were collected during the months of June-July and December 2010, respectively, following the NTC 756 Standard.

Samples were taken independently from two plots, each of one hectare (100 m by 100 m) in the Finca los deseos (Plane coordinates: N 1014988 - W1527263), located in the village of el Carmen, in the township of Mandinguilla and the Finca Nueva Esperanza (Plane coordinates: N 1008903 - W1524140) located in the township of Higo amarillo. A total of sixteen sites were sampled corresponding to a 20m X 20m mesh (Figure 10), taking fruit from lines 20, 40, 60 and 80m for a total of 16 trees.

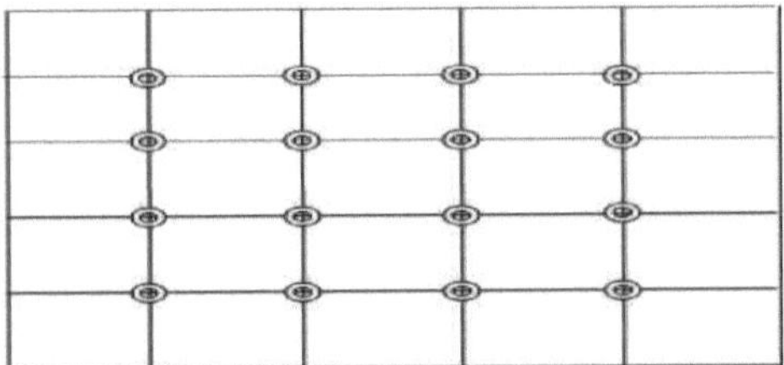

Figure 10. Schematic of fruit sampling.

From each tree, 50 fruits of different stages of maturity were collected, according to the table established in NTC 4086.

The fruits were collected in plastic baskets, washed with potable water to remove dirt and taken to the pilot plant of the agribusiness program of the Universidad Popular del Cesar to select whole, firm, fresh, healthy oranges, free of visible foreign matter, free of black spots, marked bruises, insects, external humidity, free of insect attack, free of damage due to pronounced temperature changes, free of foreign smell and taste, in accordance with the NTC 1268 standard.

5.3 Physicochemical analysis of Valencian oranges.

The physicochemical analyses were performed according to Table 7.

Table 7. Physical analysis of Valencian oranges.

Physical Characterization of the Fruit	
Parameter	**Method**
Color determination	NTC 4086

Total weight	Gravimetry
Juice, shell and seed content	Gravimetric classification
Diameter.	Direct measurement

Source: Pectin research seedbed

The orange juice was removed using a manual juicer and the residue containing exocarp, membranes and vesicles was washed with abundant water to remove solid impurities and solubilize sugars. This material was cut into 1 x 1 cm pieces (see image 1) and stored at 4°C.

Image 1. Orange peels cut into pieces.

The juice and peels were subjected to the following analyses (Table 8)

Table 8. Physicochemical analysis of orange juice and peels

Physicochemical analysis of the juice.	
Parameter	**Method**
PH	NTC 4592
Determination of titratable acidity	NTC 4086 and NTC 4623
Determination of total soluble solids.	NTC 4086 and NTC 4624
Determination of maturity indexes	Relationship between acidity and total soluble solids.
Physicochemical analysis of the hulls.	
Moisture content	A.O.A.C
Percentage of ash	Calcination at 550 C°
Determination of fat content.	NTC 668

Determination of crude fiber content	NTC 668

Source: Pectin and polysaccharides research group.
A completely randomized design was used, comparing the contents at different stages of maturity using Duncan's multiple comparison test at 95% probability.

According to the results obtained in the physicochemical analysis of the orange, the best characteristics were obtained at maturity stage 2, so the essential oil and pectin extraction tests were carried out at this stage.

5.4 Extraction of Essential Oils.

5.4.1 Steam entrainment extraction.

The method used for the extraction of essential oils was the steam extraction method; for this, 50 g of the sample to be extracted (flavedo, albedo and remaining juice vesicles) were chopped and then deposited in a volumetric ball with lateral detachment with a 1/3 shell-water distillate ratio (150 ml). Then an assembly was assembled as shown in image 2 (some glass beads were added).

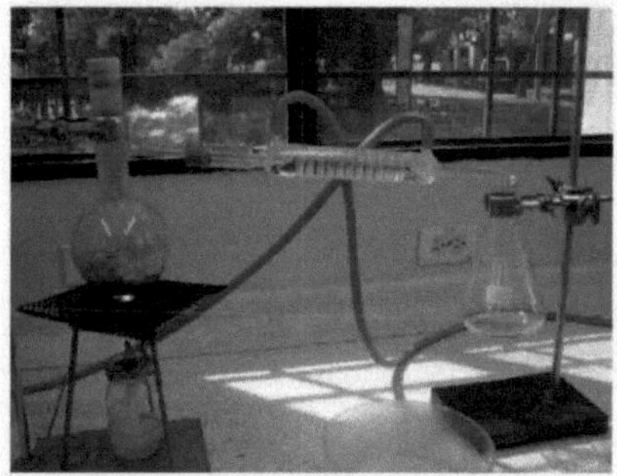

Image 2. Extraction of essential oils by vapor entrainment.

The balloon was heated with the water using the burner; the steam formed entrained the oils which then condensed to be collected in another vessel.
Heating was maintained until there was no more oil release.
The condensate was then evaporated in a water bath until the oil remained at the bottom of the container, after which the yield of essential oil obtained was quantified by weight difference.

Since the yields obtained were too low (<0.1%) and the separation of the essential oil was impossible because it remained adhered to the balloon, it was decided to use another extraction method. This consisted of Clevenger with microwaves.

5.4.2 Extraction by the hydrodistillation method with microwave-assisted Clevenger device.

The samples were placed in a 2.5 L balloon that was introduced into a 2.45 GHz MARS microwave equipment at atmospheric pressure, the temperature of which was controlled by RPT 300 plus. The balloon was coupled to a Clevenger device cooled with ethanol at -5°C and a magnetic stirring system. Extractions were performed in triplicate at a power of 600 W, time of 10 min, with 150 mL of water or in the absence of solvent. The collected essential oil was dried with anhydrous sodium sulfate, weighed and analyzed.

5.5 Analysis of the essential oil obtained.

The characterization of the oil was carried out in the environmental catalysis group of the engineering faculty of the University of Antioquia. An *Agilent Technologies Series GC System* gas chromatograph (Palo Alto, USA), coupled to an *Agilent Technologies* mass selective detector, equipped with a *split/splitless* injection port (1:30 *split*) and an *HP ChemStation* 1.05 data system was used. A DB-5MS capillary column (J *&* *W Scientific, Folsom*, USA) with 5% phenyl-polymethylsiloxane stationary phase (60 m x 0.25 mm, D.l. x 0.25 pm, df) was used for separation of the mixtures. The oven temperature was programmed from 45° C (5 min) to 150° C (2 min) at 4OC/min, then increased to 250OC (5 min) at 5OC/min. Finally, the temperature was increased at 10OC/min to 275OC (15 min). The ionization chamber and transfer line temperatures were 230 and 285OC, respectively. The carrier gas was helium (99.995%). The reconstructed mass and ion current spectra were obtained by automatic frequency scanning (full *sean*), at 4.75 scan s^{-1} , in the mass range *m/z* 30-450.

The extractions were performed in triplicate and the analysis of the essential oil by the hydrodistillation method with Clevenger device was performed in the environmental catalysis laboratory at the UdeA. See image 3.

Extraction of essential oils by the hydrodistillation method with Clevenger device.

5.6 Pectin Extraction

5.6.1 Enzyme Inactivation

Inactivation of pectinolytic enzymes from orange peels.

To inactivate the enzymes, 250 g of the stored sample were taken in a 1000 ml Beaker with a ratio of peel to distilled water of 1-3 (750 ml of distilled water), then the sample was heated to boiling point (98 - 99° C) (see image 4), remaining at this temperature for 10 minutes. At the end of heating, the peels were immersed in cold water to prevent degradation of the pectins by heat. The inactivation water was discarded.

5.6.1 Acid hydrolysis

The same amount of acidified water initially used was added to the solid material, i.e., a plant material - water ratio of 1:3, and the solution was heated to boiling temperature (see image 5) for the time determined in the treatments (Table 9) and stirred constantly to avoid sedimentation and degradation of the bagasse.

Image 5. Acid hydrolysis of orange peel.

Table 9. Treatments for pectin extraction.

TREATMENT	P H	TIME(min)	ACID O
T1	2, 0	30	HCL
T2	3, 0	50	HCL
T3	2, 0	30	HNO3
T4	3, 0	50	HNO3
T5	2, 0	50	HCL
T6	3, 0	30	HCL
T7	2, 0	50	HNO3
T8	3, 0	30	HNO3

Source: Pectin research seedbed

Table 12 shows the experimental design used in the acid hydrolysis of the pectin extraction process, 8 treatments were applied, involving 3 independent variables with two levels: type of acid (HNO3 and HCl), pH (2 and 3), and hydrolysis time (30 and 50 minutes); a standard temperature of 98° C or boiling temperature was used for all treatments.

5.6.2 Filtration

The solution resulting from the hydrolysis was cooled and filtered using linen cloth until the bagasse was very dry (see image 6).

Filtration of bagasse after hydrolysis.

5.6.3 Precipitation

Precipitation was performed with commercial grade ethanol (approx. 96% w/w).

A volume of 80% ethanol per volume of extracted solution was used for precipitation.

Ethanol was added to the solution slowly with constant stirring and allowed to stand for one hour (see Figure 7).

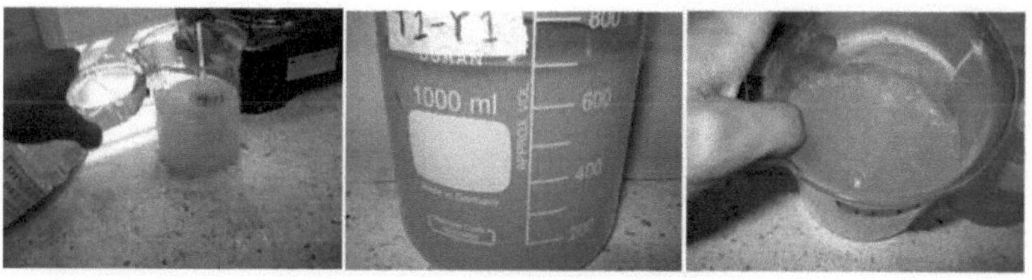

Image 7. Precipitation of pectin with 96% w/w commercial ethanol.

5.6.4 Gel filtration

After one hour of precipitation a biphasic solution was formed. The upper phase was characterized by its gelatinous texture composed mainly of pectin, the lower phase consisted of ethanol, traces of pectin and other pectin-soluble compounds.

A new filtration process was carried out, separating the supernatant that is the Pectin from the solution with a linen cloth (see image 8).

Separation of pectin by filtration of the alcoholic solution.

5.6.5 Drying

The filtered pectin was placed on watch glasses (see picture 9) and dried in a natural convection oven at 45 +/- 5°C. (Memmert) drying took approximately 72 hours, weighed to constant weight.

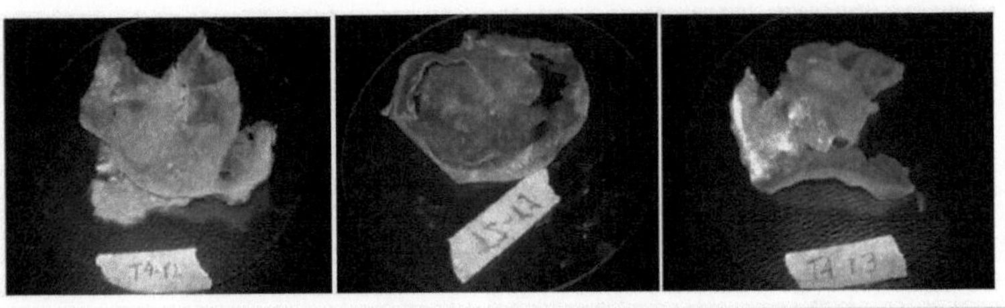

Image 9. Dried pectin

5.6.6 Shredding

In order to reduce the particle size of the pectin, a KIA electric mill was used to pulverize the sample.

The experimental design used is completely randomized, comparing treatments according to the acid used, using Duncan's multiple comparison tests at 95% probability.

5.7 CHARACTERIZATION OF THE PECTIN OBTAINED

Yield in weight: It was determined by the ratio of the weight of pectin obtained (w2) and the weight of processed peel (w1), using the following expression:

$$Yield\ (\%) = \frac{w2}{w1} x100 \quad \frac{w2}{w1} x100$$

Degree of Esterification: The determination of the degree of esterification was developed according to the technique described below (AOAC. 1980; A. M. Bochek, 2001).

- They were weighed (0.2 g) of sample and added to a container containing 2 measured ethanol.
- Twenty mL of distilled water was added at 40^0 C and kept under constant stirring on a magnetic stirrer for a period of 2 hours.
- The resulting solution was titrated with 0.1 N NaOH in the presence of phenolphthalein until it turned pink. The volume spent in ml of NaOH for this first titration was measured, the amount of NaOH used in this first titration was denominated "A" and the number of free carboxyl groups (%) was calculated with the following equation:
  Where:
  a = amount of sample weighed
  N_{NaOH} = normality of the alkaline solution spent in the titration.
  A = volume of alkaline solution spent in the titration.
- After determination of the free carboxyl groups, 10 mL of the 0.1 N NaOH solution was added to the neutralized galacturonic acid (pectin) sample. It was capped and left in agitation at room

$$Kf = \frac{NNaOH * A}{a} \times 100$$

temperature for 2 hours to saponify the esterified carboxyl groups of the polymer.

- Then 10 ml of 0.1 N HCl was added, the excess HCl was titrated with 0.1N NaOH. This amount was labeled "B".
- The number of carboxyl groups was calculated from the volume of NaOH spent for the second titration:
  Where:
  a = amount of pectin (sample) weighed
  N_{NaOH} = normality of the spent alkaline solution in the second titration.

$$Ke = \frac{NNaOH * B}{a} \times 100$$

B = volume of alkaline solution spent in the second titration.

- The total number of carboxyl groups (%), K_t, is equal to the sum of the free and esterified groups:
  $Kt = Kf + Ke$
- The number of methoxylated groups in galacturonic acid (pectin) was calculated by:

$$K_{MeO} \text{ ¿ } \frac{100\, ED\, x\, 31}{176 + ED\, X\, 14}$$

- Expressing the degree of esterification, DE, as fractions of unity.

$$\text{ED ¿ } \frac{Ke}{Kt} x\,100 = \frac{Kt * Kf}{Kt} x\,100 = \frac{1 - Kf}{Kt} x\,100$$

56.328 mg of galacturonic acid (AGA) 97% and monohydrate were weighed, dissolved with 0.25 ml of NaOH 1 N and diluted to 500 ml with distilled water. Thus a standard of 100 pg of AGA per mL was obtained. From which the elements to prepare the curve were prepared in 50 mL buckets according to Table 10.

Galacturonic acid content: This test was performed on the treatments with the highest pectin yield (T3 and T1 from the pectin extracted from the orange samples from nueva esperanza; T1 and T7 from the pectin extracted from the orange samples from los deseos). Three replicates were performed for each treatment.

The galacturonic acid (AGA) content was determined by the 3,5-dimethyldiphenol colorimetric method, following the procedure below:

Table 10: Preparation of the calibration curve.

- **ation of the calibration curve.**

Elements	1	2	3	4	5	6	7	8
AGA100 µg /mL	0	2,5	5	7,5	10	20	40	50
Distilled water (ml)	50	47,5	45	42,5	40	30	10	0
[AGA µg /ml]	0	5	10	15	20	40	80	100

Sample analysis.

A 10 mg sample of pectin (dry) was weighed into a 50 mL beaker to which 4 mL of concentrated

sulfuric acid was added, stirred gently and continuously. 1 mL of distilled water was added dropwise, and after 5 min more of stirring another 1 mL of distilled water was added dropwise. Stirring was continued for 30 min to complete the acid hydrolysis of the polysaccharides present in the sample. Once the sample was hydrolyzed, it was filtered over glass fiber, whose residues and subsequent washes (with distilled water) were collected in a 50 mL volumetric balloon, with a subsequent dilution to the final volume (50 mL).
These extracts were stored at 4^0 C for 24 hours, until the corresponding procedures were performed.

Color development
From an aliquot of 1.4 ml of each standard (all the concentrations from 0 to 100 or 8 tubes are needed and with the one of zero the equipment was calibrated) taken with micropipette to beaker of 50 ml, were added 8.4 ml of Sulfuric Acid/Tetraborate solution (0.0125 M Sodium Tetraborate in concentrated Sulfuric Acid) were added and taken to an ice bath and carefully mixed using a vortex (tube shaker) at moderate speed with intermittent stops to ensure complete mixing.
The beakers were then heated in a boiling water bath for 5 minutes and immediately placed in ice water to cool them. To the beakers 0.14ml of 0.15 % m-Hydroxydiphenyl (prepared in 0.5% NaOH) was added, vortex mixing was performed and they were subsequently kept at rest at room temperature for 15 min before carrying out the final reading of Absorbance at 520 *nm* in the spectrophotometer.

pH: It is determined by using a pH meter, calibrating with the pH of a 1% solution of pectin in distilled water.

Free acidity: It is determined by volumetric titration, preparing a solution of 0.5 g of pectin in one liter of distilled water, neutralizing 100 ml of solution with 0.1N sodium hydroxide, using phenolphthalein as an indicator.

$$Acidez\ Libre = \frac{ml.\,de\,NaOH * Normalidad}{peso\ muestra(g)}$$

Ash percentage: It was determined by direct incineration. We weighed 1 ± 0.5 grams of sample (w1) by incinerating it in a muffle at a temperature between 550 +/- 50°C for four hours, then weighing the ash obtained (w2). The percentage of ash was calculated according to the following expression:

$$\frac{W2}{W1} x100 \quad \frac{w2}{w1} x100$$

Ash (%) =

Moisture percentage: It is determined by using an oven, at a temperature of 40°C. The sample is initially weighed (w1) and dried to constant weight (w2). The moisture percentage is calculated according to the following expression:

$$Humidity\ (\%) = \frac{W1 - W2}{W1} x100$$

Equivalent weight: To determine the equivalent weight, the method described by R. M. McCready -in Methods in Food Analyses, Joslyn, 1970- was applied as follows: Duplicate samples of 0.5 g of pectin were weighed on analytical balance and transferred to a 250 ml. beaker. The samples were then moistened with 5 ml. of ethanol and 100 ml. Of boiled and cooled distilled water was added. The pectin was dissolved with the aid of a magnetic stirrer and slowly titrated with 0.1N sodium hydroxide to pH 7.5. The end point was held for 30 seconds and observed with the aid of a potentiometer. The titrated solution was kept for determination of % methoxyl.

$$peso\,equivalente = \frac{peso\,muestra(mg)}{ml.\,de\,NaOH * Normalidad}$$

Gelation potential test: Jellies were prepared with the extracted pectin by mixing water and sugar until obtaining 20° Bx. The pH was adjusted (2.2, 2.8 and 3.4), which was kept constant throughout the process by addition of citric acid, pectin was added at concentrations of 0.5 and 1.0% (the final concentration of the formulation) and was brought to different oBrix (55, 65, 80%) with addition of commercial sugar (Fiszman S.M., 1984).
A sensory evaluation was performed to assess range, consistency and structure as follows:

a. **Range**
No gelation (Liquid): the gel remains liquid or highly viscous (L) **Good gelation:** good, consistent gelation (G)
Pre-freezing: partially shattered gel, soft and viscous texture (P)

b. **Consistency:** Describes the density, firmness and viscosity of the sample. It is observed by pressing with a glass rod. This evaluation is carried out taking into account the following evaluation criteria (see table 11).

Table 11. Evaluation criteria for the consistency of pectin gels.

Consistency				
Liquid o	Very soft	Bland or	Moderately hard	Dur o
0	1	2	3	4

Source: Pectin Research Seminary

c. **Structure:** describes the bonding and homogeneity of the sample. It is observed on the surface of a shattered gel.
- Rough and brittle surface: non homogeneous gel.
- Smooth surface: homogeneous gel.

Degree of gelation: These are the units of weight of sugar with respect to a unit of weight of pectin, at pH 3 and 65% of soluble solids. In 10 glasses 0.1 to 1.0 g of pectin were placed, adding 50 ml of a citric acid solution of pH 3, heating until dissolution, adding enough sucrose to each glass so that each one results with 65% of soluble solids, completing the volume to 100 mi with the acid solution. It is left to cool for 24 hours to gel. The assumed grade is correct if at the moment of inverting the beakers, their content is only reduced by 10% of their initial height (Gierselmer K., 1997).

Time and speed of gelation: The determination of the time and speed of gelation, are two parameters that allow us to know in which types of products the obtained pectin can be used. A jelly is prepared with the extracted pectin by mixing water and sugar until 20° Bx is obtained. The pH is adjusted to 3.0, which will be kept constant throughout the process by the addition of 0.05 N HCl, the pectin is added (0.5% the final concentration of the formulation) and it is brought to 65OBx with the addition of sugar. The prepared jelly is poured into a beaker, placed in a water bath at 30OC and a timer is set. Both the glass and the jelly can be observed through the walls of the glass. The contents of the beaker are given a light, gentle swirl at intervals. At first only particles of jelly will be seen to rotate in the direction in which the glass was moved. As soon as curdling begins, it will be seen that the particles located in the lower part of the glass first rotate in the same direction and then acquire a movement in the opposite direction. The stopwatch stops when the last portion of jelly at the top of the glass is seen to move in the opposite direction to that in which it was imprinted on the glass. The time that marks the stopwatch indicates the speed of gelation, which is applicable to both pectin and jelly (Fiszman, 1984).

Chapter 6

6. RESULTS AND DISCUSSION

6.1. Physicochemical analysis of valence orange.

6.1.1. Color determination

The raw material is classified into six stages of maturity, according to the following color scale established in the NTC 4086 standard for Valencia oranges produced below 700 m.a.s.l. (image 10); image 11 shows the stages of maturity found in the lots under study.

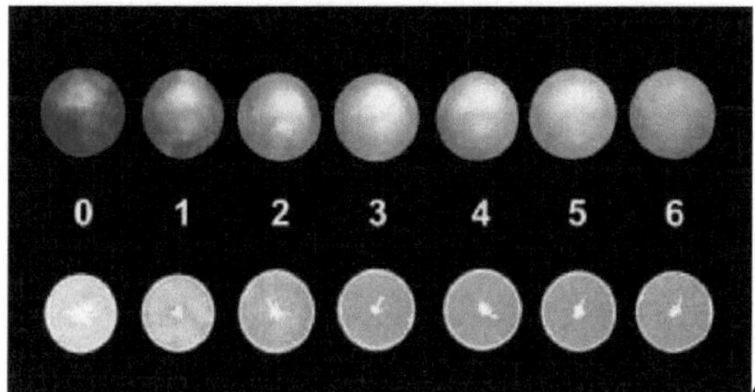

Color chart of Valencia orange according to the standard

Color chart of Valencia orange found in the lots under study.

Because growers harvest their oranges at maturity stages 2 to 3 maximum, it was difficult to find fruit at maturity stages 4 and above.

The oranges found in the lots presented the characteristic shape of the Valencia orange. They presented calyx, were healthy (free of insect attacks and/or diseases, which could degrade the internal quality of the fruit), were free of abnormal external humidity, were free of any foreign smell and/or taste, presented a fresh and firm appearance.

The following physicochemical analyses were carried out on the oranges, both on the whole fruit and on the juice and peels. The results are presented first for the Los deseos farm and then for the Nueva Esperanza farm.

6.1.2. Physical characterization of the fruit - Los Deseos farm

The physical analyses of the Valencian oranges from the Los deseos farm are shown below (Table 12). A characterization was made by maturity stage.

Table 12. Physical characteristics of oranges from the Los deseos farm.

⁰Mature z	Weight (g)	0 (mm)	% Juice	% Shell	% Seed
1	158.9±43.73b	67.7±6.6b	50.9±6.2b	46.82±5.2	2.64±0.35
2	179.0±58.8C	69.9±8.6b	50.4±12 .1b	46.42±6.3	1.96±0.15
3	178.0±54.1bc	69.7±6.6b	50.1±12 .7b	42.53±4.1	2.19±0.15
4	221.5±52.1bcd	75.9±7.0b	48.4±3. 9b	41.99±5.3	1.53±0.25
5	205.7±95.8d	73.1±12 .1b	50.8±0.2b	52.02±7.6	1.31±0.56

6.1.3. Physicochemical characterization of the juice - Los Deseos farm.

The juice extracted from the oranges was subjected to the established physicochemical analyses and the results are shown in the following table (Table 13).

Table 13. Physicochemical characteristics of orange juice from the Los Deseos farm.

oMadure z	oBrix	PH	Acidity (%)	Maturity index oBrix/Acidity index
1	10.4±0.7b	3.8±0.5C	0.73±0.2b	14.28
2	10.9±1.2C	4.0±0.5Cd	0.56±0.1b	19.4
3	10.7±1.3bC	4.0±0.4Cd	0.56±0.2b	19.12
4	12.0±1.8C	4.0±0.4d	0.51±0.02b	23.52
5	12.5±1.2d	4.2±0.2e	0.54±0.1b	23.14

In all the variables analyzed in Table 12 and 13, ANOVA showed statistically significant differences between the means at a confidence level of 95 %. Using the Multiple Range Test, it was found that the means are significantly different from each other and that indices with the same letter do not show significant differences.

In the case of diameter, no differences were found in maturity grades 1 to 5. In terms of juice percentage, all maturity grades showed similarity. In the^{0} Brix there is similarity in maturity color 1 to 4, which indicates that the fruit could be harvested from maturity color 1.

The average juice percentage is 49.61% and oBrix is 10.8, very close to the quality criteria required by the industry of 50% and 10.5, respectively. Acidity is between 0.5 and 0.7 in almost all points.

As for pH, there is similarity in color between 2 and 4 and it increases as the degree of maturity increases. In acidity, there are no significant differences in the maturity grades, it decreases as the fruit ripens and the pH increases. The maturity index also increases as fruit maturity increases. However, it is observed that in color 1, the % acidity is at the upper limit required by the industry and that it would be important that the harvest be carried out at the maximum maturity color 2, since from this grade, the oBrix increases substantially, reaching values close to 12.5 in color 5.
This is corroborated by checking the maturity index, which is higher than 15 at the same degree of maturity. This is favorable since it allows the fruit to be handled and transported without any deterioration until its final destination.

6.1.4. Physical characterization of fruit - Nueva Esperanza farm

The physical analyses of the Valencian oranges from the Nueva Esperanza farm are shown below (Table 14). The characterization was done by stage of maturity.

Table 14. Physical characteristics of the orange from the Nueva Esperanza farm.

0 Maturity	Weight (g)	0 (mm)	% juice	% shell	% seed
1	210±46. 5	73.6±6. 0	55.2± 5.7	44.43± 5.6	1.52±0. 46
2	192.2±4 2.8	72.0±6. 3	54.3± 5.5	44.75± 4.5	1.94±0. 77
3	173.6±2 2	70.1±4. 4	55.2± 4.8	42.34± 4.1	2.42±0. 72
4	163.7±6 8.1	71.4±1 4.4	47.7± 6.9	43.13± 5.1	1.34±0. 35
5	215.1±5 7.6	75.3±9. 5	47.8± 2.0	48.07± 7.2	2.16±0. 53

Physical and chemical characterization of juice - Nueva Esperanza farm

The juice extracted from the oranges was subjected to the established physicochemical analyses and the results are shown below (Table 15).

Table 15. Physicochemical characteristics of orange juice from the farm.
New Hope.

0M d	oBrix	pH	Acidity (%)	maturity index oBriXAcidity index

1	10±0. 7	4.1± 0.4	0.58±0.3	19.5±8.6
2	10.1± 0.8	4.0± 0.2	0.55±0.1	19.1±5.9
3	10,7± 0.6	4.0± 0.1	0.51±0.1	20.6±3.5
4	9.2±0. 4	3.7± 0.3	0.5±0.2	22,0±2,1
5	11.5± 0.4	3.7± 0.1	0.49±0.0	22,4±3,0

The juice content of all stages of maturity exceeds 40%, the soluble solids vary between 9.2 and 11.5, the acidity is less than 0.58 and the maturity index exceeds 19.1, exceeding all the values established by standard 4086, which leads to the conclusion that from color one onwards all oranges meet the specific requirements of the standard and can be consumed fresh, contrary to what is established by NTC4086, which suggests that the best organoleptic characteristics for the fresh and processed market are from color 3 onwards.

It is also observed that from color two, this orange meets the requirements to be used in the industry, such as a % juice higher than 50%,^{0} Brix higher than 10.5, acidity between 0.5 and 0.7 and a maturity index higher than 15%. Ten percent of the samples analyzed have a juice percentage lower than 40 %, although they have a C size and diameter greater than 63 mm.

After the pertinent analyses of the whole fruit from the two farms and the juice extracted from them, the peels were subjected to physicochemical analysis and essential oil and pectin extractions.

6.1.5. Physicochemical characterization of the hulls.

The peels, the object of study in this project, yielded the following results (see Table 16) from the physicochemical analyses established in the methodology.
The results of the physicochemical analysis of orange peels from the two farms are shown below.

Table 16. Percentage of moisture, ash, fiber and ethereal extract of orange peels from the two batches.

Lot	% Humidity	% Ashes	% Crude fiber	% ethereal extract
Wishes(L1)	70.40±1.170	1.183±0.270	10.12±0.62	0.27 ± 0.026
New hope(L2)	72.00±1.170	1.528±0.674	9.82 ± 0.54	0.26 ± 0.046

We observed that both the percentage of humidity and the percentage of ashes in the peels of the Valencia orange is lower on the Los Deseos farm, perhaps due to the management conditions of the crop, since the Los Deseos farm does not have an irrigation system or adequate fertilization of the soil, while on the Nueva Esperanza farm the crop is more technified, with adequate control of irrigation and soil fertility.

The fiber percentage was 10.12% with a standard deviation of 0.62 on the Los Deseos farm and 9.82% with a standard deviation of 0.54 on the Nueva Esperanza farm; these values are below the value of 13% reported by Demain and Solomon, but nevertheless are very close. As for fat content or ethereal extract, the percentages obtained: 0.26 and 0.27%, were well below the value reported by these same authors, which was 3.4%.

The physicochemical analyses carried out on Valencian oranges in their different degrees of maturity lead us to conclude that the best stage for harvesting is maturity stage 2, because at this point the orange has all the quality characteristics that the specific regulations require for it to be industrialized; hence it is necessary to test the extraction of pectin and essential oils from the peels of the oranges at this stage of maturity.

6.2. PECTIN EXTRACTION

6.2.1. PECTIN YIELDS

6.2.1.1. END OF WISHES (L1)

The yields obtained from fresh Valencian orange peels from the Los Deseos farm are shown below (Table 17); one column shows the yields with HCl and the other with HNO_3.

Table 17. Pectin yield of orange pectin from the Los Desejos farm.

pH	**time**	**T**	**Yield (%) HCl**		**T**	**Yield (%)** HNO3	
2.0	50 MIN	T5	9,620 ±	0,764 a	T7	7,780 ±0,904	a
	30 MIN	T1	8,968 ±	1,409 a	T3	7,079 ±1,276	a b
3.0	50 MIN	T2	5,652 ±	0,735 b	T4	5,705 ±0,648	b c
	30MIN	T6	4,183±	0,497 b	T8	4,823 ±0,372	c

The best treatments obtained in which HCl was used as hydrolyzing agent were: T5 with a yield of 9.620% with a standard deviation of 0.764 and T1 with a yield of 8.968% with a standard deviation of 1.409; which did not present statistically significant differences at a 95.0% confidence level; with which we can conclude that although treatment 5 presents a higher yield than T1, this difference is not statistically significant.

significant, so treatment 1 becomes the best as it requires less hydrolysis time (30 minutes). The

lowest yields were achieved at pH 3, thus we see how treatment 6 (T6) had a yield of 4.183% with standard deviation of 0.497. We observe then that the yields increased as the hydrolysis time increased and decreased as the pH increased.
Regarding the yields obtained with nitric acid, the best treatments were T7 (pH2 and 50 minutes of hydrolysis) with a yield of 7.78033% with a standard deviation of 0.904 and T3 (pH2 and 30 minutes of hydrolysis) with a yield of 7.079% with a standard deviation of 1.276, which presented a slight statistically significant difference with a level of 95% confidence; with which it is concluded that the best treatment was T7. The lowest yields with HNO3 were obtained with high pH values and short hydrolysis time, as shown by treatment 8 (T8) with a yield of 4.823% and standard deviation of 0.372. Comparing the best treatment using HCl (T1) and HNO3 (T7), it is concluded that the best treatment was T1, since it presented a higher yield; however, the yields of these two treatments did not show statistically significant differences at a 95% confidence level, as shown in Table 18, where 5 homogeneous groups have been identified according to the alignment of the X's in columns. There are no statistically significant differences between those levels that share the same X's column.
Table 18. Multiple Range Tests T1-T7 (Method: 95.0 LSD percentages)

	Cases	*I measured*	*Groups Homogeneous*
T6	3	4,18267	X
T8	3	4,823	X
T2	3	5,652	XX
T4	3	5,705	XX
T3	3	7,079	XX
T7	**3**	**7,78033**	**XX**
T1	**3**	**8,96767**	**XX**
T5	3	9,62	X

The best treatment performed on the pectin from the Los deseos farm was T1 because compared to the other best treatment T7 with HNO3 (pH 2.0 and 50 min.), T1 requires a shorter hydrolysis time (30 min.); moreover, a prolonged time has repercussions on the quality of the pectin, especially in the degree of esterification.

6.1.5.1. NUEVA ESPERANZA FARM (L2)

The yields obtained from fresh Valencian orange peels from the Nueva Esperanza farm are shown below (Table 19); one column shows the yields with HCl and the other with HNO_3 .

Table 19. Orange pectin yield from the nueva esperanza farm.

pH	time	T	Yield (%) HCI	T	Yield (%) HNO3
2.0	50 MIN	T5	6,001 ± 0,295^{a}	T7	9,614 ± 1,064^{a}
	30 MIN	T1	7,385± 0,112^{b}	T3	9,968 ± 1,114^{a}
3.0	50 MIN	T2	2,909± 0,163^{c}	T4	3,926 ± 0,199^{b}
	30 MIN	T6	1,907± 0,451^{d}	T8	3,109 ± 0,592^{b}

Analyzed these results, we can say that the best treatments were those that used HNO3 as hydrolyzing agent, these were: T3 with a yield of 9.968% with a standard deviation of 1.114 and T7 with a yield of 9.614% with a standard deviation of 1.064; which did not present statistically significant differences with a level of 95.0% confidence; with this we conclude that the best treatment with HNO3 was T3, for the shortest hydrolysis time and the highest yield. As for the lowest yields, we observed how the high pH and short extraction times affect a low pectin yield, thus, in treatment 8 (T8: HNO3, pH 3.0 and 30 min) a yield of 3.109% was obtained with a standard deviation of 0.592 and when increasing the time to 50 minutes the result was 3.926% with a standard deviation of 0.199. Regarding the yields obtained with HCl, the best treatments were T1 (pH 2 and 30 minutes of hydrolysis) with a yield of 7.385% with a standard deviation of 0.112 and T5 (pH 2 and 50 minutes of hydrolysis) with a yield of 6.001% with a standard deviation of 0.295, which presented statistically significant differences with a level of 95% confidence; with which it is concluded that treatment 1 is the best due to the higher yield and shorter hydrolysis time. Regarding the yields with HCl at pH 3.0, the same behavior is observed as with HNO3, i.e., as the pH increases and the extraction time decreases, the yields decrease; thus we see that with a time of 30 minutes (T6) a yield of 1.907% was obtained with a standard deviation of 0.451 and with 50 minutes a yield of 2.909% with a standard deviation of 0.163.

Comparing the best treatment using HCl (T1) and HNO3 (T3), it is concluded that the best treatment was T3, since it presented a higher yield, in addition, the yields of these two treatments showed statistically significant differences with a 95% confidence level, as shown in Table 20; where 5 homogeneous groups have been identified according to the alignment of the X's in columns. There are statistically significant differences between those levels that do not share the same X's column.

Table 20. T1-T3 Multiple Range Tests (Method: 95.0 LSD percentages)

	Case s	*I measured*	*Homogeneous Groups*
T6	2	1,907	X
T2	2	2,9085	XX
T8	3	3,109	XX
T4	2	3,9255	X
T5	3	6,001	X
T1	**2**	**7,385**	**X**
T7	3	9,61367	X
T3	**3**	**9,96767**	**X**

It can be concluded that the highest percentages of yield were obtained in the Nueva Esperanza farm (L2), T3 = 9.968 ± 1.114 % with nitric acid as extraction agent and with hydrochloric acid the highest yield was T1 = 7.385 ± 0.112%; compared to Los Deseos (L1) where the highest yield was T5 = 9.620 ± 0.764% with hydrochloric acid as extraction agent and with nitric acid the highest yield was T7 = 7.780 ± 0.904%. We see then how the action of the extraction agents in the two farms were different and opposite.

Comparing the best treatment of each farm (T3L2 and T5L1), we observed that there are no statistically significant differences at a 95% confidence level as shown in Table 21.

Table 21. Multiple testing of T3L2 and T5L1 ranges.

	Case s	I measured	Homogeneous Groups
T5L2	3	6,001	X
T3L1	3	7,079	XX
T1L2	2	7,385	XXX
T7L1	3	7,78033	XX
T1L1	3	8,96767	XX
T7L2	3	9,61367	X
T5L1	**3**	**9,62**	**X**
T3L2	**3**	**9,96767**	**X**

We conclude then that the best treatment between the two farms was T5 (HCl, pH 2.0 and 30 MIN.) of the Los deseos farm, since the hydrolysis was performed with HCl, an extraction agent better recommended than HNO3, since the pectin will be used in the food industry and for human consumption; furthermore, there was no significant difference between these two treatments.

6.2.2. CHEMICAL AND PHYSICAL CHARACTERISTICS OF THE OBTAINED PECTIN

6.2.3. DETERMINATION OF THE DEGREE OF ESTERIFICATION AND METHOXYLATION.

Degree of esterification and methoxylation of orange peel pectin from the Nueva Esperanza farm.

Tables 22 and 23 below show the results obtained for the degree of esterification and methoxylation, respectively, of the pectin extracted from the Nueva Esperanza farm.

Table 22. Degree of esterification of pectin from the Nueva Esperanza farm.

Treatment	A* value	B* value	Degree of esterification(%ED)			%ED Average
			Test 1	Essay 2	Essay 3	
	1,30	4,50	68.586			
T1	1,20	4,40		69.571		69.479 ± 0.85^{bc}
	1,15	4,40			70.279	
	1,30	4,60	68.966			
T2	1,20	4,40		69.571		69.370 ± 0.35^{bc}
	1,20	4,40			69.571	
	1,15	4,50	70.646			
T3	1,15	4,60		71.000		71.227 ± 0.72^{a}
	1,10	4,70			72.034	
	1,25	5,20	71.620			
T4	1,20	4,90		71.328		71.212 ± 0.48^{a}

	1,30	5,10			70.688	
T5	1,40	4,40	66.862			67.657 ±0.87 d
	1,30	4,50		68.586		
	1,35	4,40			67.522	
	1,30	4,40	68.193			
T6	1,20	5,20		72.250		70.481 ± 2.08ab
	1,20	4,80			71.000	
	1,30	4,50	68.586			
T7	1,35	4,60		68.311		68.494 ±0.16 cd
	1,30	4,50			68.586	
	1,20	4,30	69.182			
T8	1,20	4,50		69.947		69.567 ±0.38 bc
	1,20	4,40			69.571	

*The I value and the B value indicate the volumes of sodium hydroxide spent for the first and second titration.

We can observe that the pectin extracted from the Nueva Esperanza farm has a degree of esterification that goes from 67.657% to 71.227%; being defined as a medium fast pectin. Only the values obtained in treatments T3 and T4 show a slight tendency to be fast gelling pectin, remaining in the lower limit of the range to be characterized as fast gelling; and treatment T5 is in the upper limit to be considered as slow gelling pectin, with a slight tendency to be medium gelling pectin.

Table 23. Degree of Methoxylation of pectin from the Nueva Esperanza farm.

Treatment	A-value	B-value	Degree of Methoxylation(%DM)			Average %DM
			Test 1	Essay 2	Essay 3	
	1,30	4,50	12.871			
T1	1,20	4,40		13.025		13.011 ± 0.13 bc
	1,15	4,40			13.136	
	1,30	4,60	12.931			
T2	1,20	4,40		13.025		12.994 ±0.05bc
	1,20	4,40			13.025	
	1,15	4,50	13.193			
T3	1,15	4,60		13.248		13.283 ±0.11a
	1,10	4,70			13.409	
	1,25	5,20	13.344			
T4	1,20	4,90		13.299		13.281 ±0.07 a
	1,30	5,10			13.199	
	1,40	4,40	12.602			
T5	1,30	4,50		12.871		12.726 ±0.14 d
	1,35	4,40			12.705	
	1,30	4,40	12.810			
T6	1,20	5,20		13.442		13.167 ±0.32 ab
	1,20	4,80			13.248	
	1,30	4,50	12.871			
T7	1,35	4,60		12.828		12.857± 0.02 cd
	1,30	4,50			12.871	
	1,20	4,30	12.964			
T8	1,20	4,50		13.084		13.024±0.06 bc
	1,20	4,40			13.025	

Methoxylation degree values were obtained for pectin from the Nueva Esperanza farm, ranging from 12.726% to 13.283%. We observed a minimal variation in the methoxyl content.

Degree of esterification and methoxylation of orange peel pectin from Los Deseos farm.

Tables 24 and 25 show the results of the degree of esterification and methoxylation of the desires farm.

Table 24. Degree of esterification of pectin from the Los Deseos farm.

Treatment	A-value	B-value	Degree of esterification(%ED) Test 1	Essay 2	Essay 3	%ED Average
	1,1	3,1	64.810			
T1	0,8	3,1		70.487		69.231 ±3.94 b
	0,8	3,5			72.395	
	1,2	3,0	62.429			
T2	1,0	2,8		64.684		64.371 ±1.81 bc
	1,0	3,0			66.000	
	0,7	4,5	77.538			
T3	0,6	4,1		78.234		77.910 ±0.35 a
	0,6	4,0			77.957	
	1,0	3,9	70.592			
T4	0,9	3,8		71.851		68.290 ±5.11 bc
	1,2	3,0			62.429	
	1,2	2,9	61.732			
T5	1,1	3,0		64.171		62.867 ±1.23 c
	1,5	3,8			62.698	
	1,1	3,0	64.171			
T6	0,7	2,9		71.556		67.863 ± 5.22 bc
	1,2	0,8				
	1,5	3,5	61.000			
T7	1,2	3,7		66.510		64.837 ± 3.33 bc
	1,2	3,8			67.000	
	1,2	3,0	62.429			
T8	1,0	3,0		66.000		65.206 ± 2.48 bc
	1,0	3,2			67.190	

The percentages of the degree of esterification of the pectin of the Los deseos farm range from a minimum value of 62.867% to a maximum value of 77.910%. This variation, unlike the results obtained from the nueva esperanza farm, is quite wide, and shows once again the incidence of crop management (use of irrigation and fertilizers) on the results obtained for the chemical composition of pectin. In treatments T2, T5, T7 and T8, the degrees of esterification obtained were below 67%, defining this pectin as slow gelling; treatments T1, T4 and T6 are considered medium gelling pectin; treatment T3 is within the range to be considered fast gelling pectin.

We cannot assign a specific range to the pectin from the Los deseos farm, since half of the treatments yielded results that indicate slow gelling, three treatments place it in the medium-fast range, and only one treatment indicates it as a fast gelling pectin. Perhaps the abnormal conditions of the soil from which the oranges were extracted (trees located at the top of a slope and others at the bottom) as well as the non-use of irrigation systems (the soil gets wetter in the lower parts of the hills during the sporadic rains), are the cause of this wide variation.

Table 25. Degree of Methoxylation of pectin from Los Deseos farm.

Treatment	A-value	B-value	Degree of Methoxylation(%DM) Test 1	Essay 2	Essay 3	-%DM Average
	1,30	4,50	12.280			
T1	1,20	4,40		13.168		12.971 ± 0.62^{b}
	1,15	4,40			13.465	
	1,30	4,60	11.905			
T2	1,20	4,40		12.260		12.210 ±0.28^{bc}
	1,20	4,40			12.466	
	1,15	4,50	14.261			
T3	1,15	4,60		14.368		14.318 ±0.05^{a}
	1,10	4,70			14.325	
	1,25	5,20	13.184			
T4	1,20	4,90		13.380		12.823 ±0.80^{bc}
	1,30	5,10			11.905	
	1,30	4,50	11.795			
T5	1,35	4,60		12.179		11.974±0.19^{c}
	1,30	4,50			11.947	
	1,30	4,40	12.179			
T6	1,20	5,20		13.334		12.757 ±0.82^{bc}
	1,20	4,80				
	1,20	4,30	11.679			
T7	1,20	4,50		12.546		12.283 ±0.52^{bc}
	1,20	4,40			12.623	
	1,40	4,40	11.905			
T8	1,30	4,50		12.466		12.341 ± 0.39^{bc}
	1,35	4,40			12.653	

Methoxylation degree values were obtained for pectin from the Los Deseos farm ranging from 11.974% to 14.318%. We observed a slightly wider variation in the methoxyl content in relation to the results obtained from the Nueva Esperanza farm.

The following is a summary of the previous tables of the results of the degree of esterification and methoxylation of the two farms, ordered according to the best treatments and by grouping of similar values or without significant differences (see tables 26 and 27).

Table 26. Esterification and methoxylation rates of the nueva esperanza farm.

TREATMENT	METHOXYLATED GROUPS(DM)	GRADE ESTERIFICATION (ED)
T3	13.283±0.112^{a}	71.227±0.721^{a}
T4	13.281 ±0.074^{a}	71.212±0.477^{a}
T6	13.167±0.324^{ab}	70.481 ± 2.078^{ab}
T8	13.024±0.060^{bc}	69.567 ± 0.383^{bc}

T1	13.011 ±0.133^{bc}	69.479 ± 0.850^{bc}
T2	12.994±0.055^{bc}	69.370 ± 0.349^{bc}
T7	12.857±0.025^{cd}	68.494±0.159^{cd}
T5	12.726±0.136^{d}	67.657 ± 0.870^{d}

Table 27. Degree of esterification and methoxylation of the Los Deseos Farm

TREATMENT	METHOXYLATED GROUPS(DM)		GRADE ESTERIFICATION (ED)	
T3	14,318t0,054 to		77,910t0,350a	
Tl	12.971 ±0.617	b	69.231 t 3.946	b
T4	12.823 ±0.801	be	68.290t5.115	be
T6	12.757t0.817	be	67.863 t 5.222	be
T8	12,341 t0.390	be	65.206 t 2.478	be
T7	12.283t0.524	be	64.837 t 3.332	be
T2	12.210t0.284	be	64.371 t1.806	be
T5	11.974 ±0.194	e	62.867 t1.228	e

The results obtained allow concluding that all of them have more than 7% methoxyl and consequently can be considered as high methoxyl pectins (Guzman, *et al.*, 1977), a clear relationship between the extraction conditions and the methoxyl content is observed and that for each extraction pH there is an optimum time at which pectin with the highest methoxyl content is obtained and that there is an optimum pH for the pectin to present the highest methoxyl content.

Comparing the high degree of esterification with the methoxyl content, it could be concluded that there are other chemical groups involved in esterification with the carboxyl groups of polygalacturonic acid, such as amide groups or neutral sugars (arabinose, rhamnose, galactose).

It can also be concluded that there is an "optimum" time within which pectin with the highest degree of esterification is obtained at a given pH and that there is also an "optimum" pH for the extraction of pectin with the highest degree of esterification.

The degrees of esterification obtained lead us to conclude that as the extraction time of pectin increases, it tends to denature, so we observe that the longer the hydrolysis time, the lower the degree of esterification and vice versa.

Tables 28 and 29 show the influence of pH and hydrolysis time on the degree of esterification of the extracted pectins.

Table 28. Influence of hydrolysis time with HCl on the degree of esterification of the pectin obtained from the Los Deseos farm.

pH	time	T	Yield(%)HCI	Degree of esterification
2.0	50 MIN	T5	9,620 ± 0,764	65,206 ± 2.478
	30 MIN	T1	8,968 ± 1,409	69,231 ± 3.946
3.0	50 MIN	T2	5,652 ± 0,735	64,371 ± 1.806
	30 MIN	T6	4,183± 0,497	67,863 ± 5.222

Table 29. Influence of hydrolysis time with HCl on the degree of esterification of the pectin obtained from the Nueva Esperanza farm.

pH	time	T	Yield(%)HCI	Degree of esterification
2.0	50 MIN	T5	6,001 ± 0,295	67.657 ± 0.870
	30 MIN	T1	7,385± 0,112	69.479 ± 0.850
3.0	50 MIN	T2	2,909± 0,163	69.370 ± 0.349
	30 MIN	T6	1,907 ± 0,451	70.481 ± 2.078

Tables 28 and 29 show that the lower the pH and the longer the hydrolysis time, the lower the degree of esterification, and when the pH increases and the hydrolysis time decreases, the degree of esterification increases.

We can conclude from the above that although a higher pectin yield is obtained under more drastic extraction conditions (low pHs and prolonged times), this directly affects the quality of the pectin, evidenced by a lower degree of esterification.

6.2.4. DETERMINATION OF GALACTURONIC ACID CONTENT (AGA)

The calibration curve for galacturonic acid taken as a reference is shown below (see Table 30).

Table 30. Calibration curve for Galacturonic acid.

Concentration (pg /mi)	**Absorbance** (**520 nm**)
0	0
5	0,029
10	0,089
15	0,168
20	0,054

Source: Pectin and polysaccharides research group. Figure 11 presents the calibration curve for the determination of the AGA content, the concentration value of 20 pg /ml with an absorbance of 0.054 nm was omitted because

40	0,491
80	0,662
100	1,093

was an abnormal value.

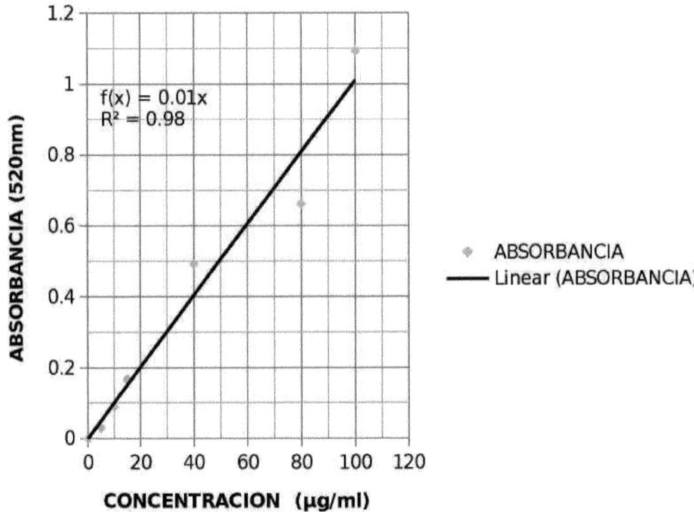

Figure 11. Calibration curve for the determination of AGA content.

Linearity obeying the Beer- Lambert law (Van Arendonk 1981), with an R^2 of approximately 0.963, is presented, and the equation governing the behavior is: y = 0.0101x

The following tables (Table 31 and Table 32) show the galacturonic acid (AGA) content of the treatments with the highest pectin yield in each of the farms.

Table 31. AGA content in pectin from the Nueva Esperanza farm.

TREATMENT	W Pectin (mg)	ABS (520 nm)	AGA Curve	%AGA	Average AGA
T3R1	10,4000	0,575	56,931	27,371	
T3R2	10,3000	0,81	80,198	38,931	35,821
T3R3	10,5000	0,873	86,436	41,160	
T1R1	10,1000	0,517	51,188	25,341	
T1R2	10,4000	0,844	83,564	40,175	29,112
T1R3	10,3000	0,454	44,950	21,821	

Table 32. AGA content in pectin from the Los Deseos farm.

TREATMENT	W Pectin (mg)	ABS (520 nm)	AGA Curve	%AGA	Average %AGA
T1R1	10,2000	0,667	66,040	32,372	
T1R2	10,4000	0,931	92,178	44,316	39,613
T1R3	10,3000	0,877	86,832	42,151	
T7R1	10,1000	0,898	88,911	44,015	
T7R2	10,4000	0,461	45,644	21,944	28,015
T7R3	10,1000	0,369	36,535	18,086	

Table 31 shows that the two treatments have almost the same extraction conditions (pH 2 and hydrolysis time 30 minutes), the difference is that treatment 1 is with HCl and treatment 2 with HNO3, then we see how a higher percentage of galacturonic acid was obtained with treatment 3 using HNO3 with a percentage of 35.821, as opposed to treatment 1 with a percentage of 29.112.

Table 32 showed a higher percentage of AGA with treatment 1 (pH 2. 30 min hydrolysis time and HCl) with a percentage of 39.613 compared to treatment 7 with a percentage of 28.015.

The purity was lower than that determined for pectins from some non-conventional sources: 66.0 g/100g in sugar beet pulp (Adomako D, 1974), 71.4 to 98.0 g/100g in sunflower heads (Michel F, 1985), 63.07 to 67.13 g/100g in soybean hulls (Miyamoto A, Chang KC, 1992) and 60.66 to71.65 g/100g in parchita hulls (Monsoor MA, Proctor A, 2001).

The low percentages of galacturonic acid in the present investigation could be due to interference from impurities such as associated neutral sugars, latex, gums and other compounds present in the structure of the Valencian orange peel that could be hydrolyzed together with the extracted pectin (Jittra *etal.*, 2005).

11.2.2.3. HUMIDITY AND ASHES

Moisture and ash data for the pectin extracted from the two farms are shown below (see Table 33).

Table 33. Moisture and pectin ash of the Nueva Esperanza and Los Deseos farms.

TREATMENT	**Nueva Esperanza farm**		**Los Deseos farm**	
	HUMIDITY, %	CENIZAS, %	HUMIDITY, %	ASH,%,% ASHES,%
T1	4,961	3,752	4,526	3,670
T2	6,090	3,891	4,796	3,545
T3	6,085	2,958	4,749	2,786
T4	5,686	3,950	4,553	3,748

T5	5,404	3,746	4,992	3,678
T6	6,329	4,223	5,222	3.506
T7	6,179	4,159	5,238	3,470
T8	5,994	3,947	4,615	3,846

From these values it can be concluded that the moisture fluctuated between 4.961 and 6.329% in the Nueva Esperanza farm and from 4.526 to 5.238% in the Los Deseos farm with a tendency to decrease in the first treatments, which seems to indicate that pectin can be obtained with a moisture around 4.5% and that due to the crushing conditions and the particle size, pectin can gain moisture up to a certain level; The moisture percentage is well below the values reported for commercial pectin and others analyzed in previous research (Giraldo, 1991), where moisture percentages of 10.81% were found for commercial pectin, 11.09% for mango pectin, 12.39% for guava pectin, 11.73% for tree tomato pectin and 8.69% for apple pectin.

The ashes fluctuated between 2.958 and 4.223% in the Nueva Esperanza farm and from 2.786 to 3.846% in the Los Deseos farm; these values are above those reported by Giraldo (Giraldo, 1991) for commercial pectin with a value of 1.34%, and above those reported by Rouse (Rouse, 1964) for the Valencia and Pineaple varieties, 1.80 and 1.72%, respectively. The ash content in pectin indicates the amount of inorganic impurities combined or occluded in the precipitate (Rouse and Knoor, 1970).

11.2.2.4. EQUIVALENT WEIGHT

The equivalent weights per treatment of the pectin extracted from the two farms are presented below (see Table 34).

TREATMENT	Equivalent Weight (mg/meq) Nueva Esperanza farm	Equivalent Weight (mg/meq) Los Deseos Farm
T1 (HCl, pH 2.0, 30')	1705,772	2603,661
T2 (HCl, pH 3.0, 50')	1696,446	2179,598
T3 (HNO3, pH 2.0, 30')	1828,209	3715,064
T4 (HNO3, pH 3.0, 50')	1676,536	2271,042
T5 (HCl, pH 2.0, 50')	1539,809	2207,615
T6 (HCl, pH 3.0, 30')	1682,721	2362,452
T7 (HNO3, pH 2.0, 50')	1576,220	1840,290
T8 (HNO3, pH 3.0, 30')	1729,463	1812,261

Table 34. Equivalent weight per treatment for the two farms

The equivalent weight value of the nueva esperanza farm is between 1539.809 and 1828.209 mg/meq and of the los deseos farm is between 1812.261 and 3715.064 mg/meq; The equivalent weights are relatively high when compared to values of 805 for lime pectin (Shrivas, 1963), 800-1000 for lemon pectin (Chaliha, 1963), 700-900 for guava pectin (Pruthi, 1960) and 600-900 for green papaya pectin (Bhatia, 1960).

6.2.2.5.

6.2.2.6. FREE ACIDITY

The results of the free acidity of the pectin extracted from the two farms are shown below (Table 35 and 36), the free acidity is expressed in milli-equivalents of free carboxyls over grams of pectin analyzed.

Table 35. Determination of free acidity of Los Deseos pectin.

TREATMENT	Normality NaOH	Volume of NaOH spent (ml)	sample weight (g)	FREE ACIDITY (meq free carboxyl/g)
T1 (HCl, pH 2.0, 30')	0,087	0,9000	0,2039	0,3841
T2 (HCl, pH 3.0, 50')	0,087	1,0667	0,2023	0,4588
T3 (HNO3, pH 2.0, 30')	0,087	0,6333	0,2047	0,2692
T4 (HNO3, pH 3.0, 50')	0,087	1,0333	0,2042	0,4403
T5 (HCl, pH 2.0, 50')	0,087	1,2667	0,2028	0,5434
T6 (HCl, pH 3.0, 30')	0,087	1,0000	0,2055	0,4233
T7 (HNO3, pH 2.0, 50')	0,087	1,3000	0,2050	0,5518
T8 (HNO3, pH 3.0, 30')	0,087	1,0667	0,2049	0,4530

Table 36. Determination of free acidity of Nueva Esperanza pectin.

TREATMENT	Normality NaOH	Volume of NaOH spent (ml)	sample weight (g)	Free acidity (meq free carboxyl/g)
T1 (HCl, pH 2.0, 30')	0,0985	1,2167	0,2043	0,5862
T2 (HCl, pH 3.0, 50')	0,0985	1,2333	0,2060	0,5895
T3 (HNO3, pH 2.0, 30')	0,0985	1,1333	0,2040	0,5470
T4 (HNO3, pH 3.0, 50')	0,0985	1,2500	0,2063	0,5965
T5 (HCl, pH 2.0, 50')	0,0985	1,3500	0,2047	0,6494
T6 (HCl, pH 3.0, 30')	0,0985	1,2333	0,2043	0,5943
T7 (HNO3, pH 2.0, 50')	0,0985	1,3167	0,2043	0,6344
T8 (HNO3, pH 3.0, 30')	0,0985	1,2000	0,2043	0,5782

The free acidity of the Los Deseos farm varies from 0.2692 to 0.5518 meq free Carboxyls/g. and in the Nueva Esperanza farm the values vary from 0.5470 to 0.6494 meq free Carboxyls/g. These results are below the values reported by Giraldo (Giraldo, 1991): 1.32 meq. for commercial pectin indicating the null residuality of the acids used (which happened with the data of the pectins used by us, the commercial ones) and very similar to the values reported for other sources: 0.99 for mango pectin, 0.6 for guava pectin, 0.54 for tree tomato pectin and 0.72 for apple pectin.

The values of free acidity of the two batches allow us to conclude that the acidity increases as the pH of the extraction medium becomes less acidic and the extraction procedure becomes more drastic (longer hydrolysis time), but it is observed that the longer hydrolysis time has a greater incidence on the acidity values; analyzing table 36, if we compare treatment 1 (T1) with treatment 6 (T6), where the agent (HCl) and the extraction time (30 minutes) are the same, but the pH varies from 2.0 to 3.0,

we find that the pH varies from 2.0 to 3.0.0 to 3.0, we realize that the acidity increased 1.36%; while keeping the extraction agent (HCl) and pH (2.0) constant and varying the times from 30 to 50 minutes, in this case T1 and T5, the variation in acidity was 9.73%.
The same occurs with nitric acid, in the first case (increasing pH only: T3 and T8), acidity increased 5.4%; and in the other case (increasing extraction time: T3 and T7), acidity increased 13.78%.
The same behavior was observed in the free acidity values of pectin from the Los Deseos farm (Table 35).

6.2.2.7. GELATION POTENTIAL TEST

Figure 12 shows the behavior of the treatments with different percentages of pectin at different pH and^0 Brix. It is observed that 55^0 Brix is the limit of gelation, since at this value, a very small syneresis (water leakage) begins to form and towards lower values there is no gelation and a viscous solution is appreciated. Between 55 and 65 oBrix, there is a good gelation, which is even maintained above 65 oBrix and from 70 oBrix, pregelation appears with a partially shattered gel, soft and viscous texture. This effect can be observed in image 12, which shows the photographic record at different pH, different oBrix and different percentages of pectin, and Table 37 shows the data on consistency, structure and gelation for all treatments.

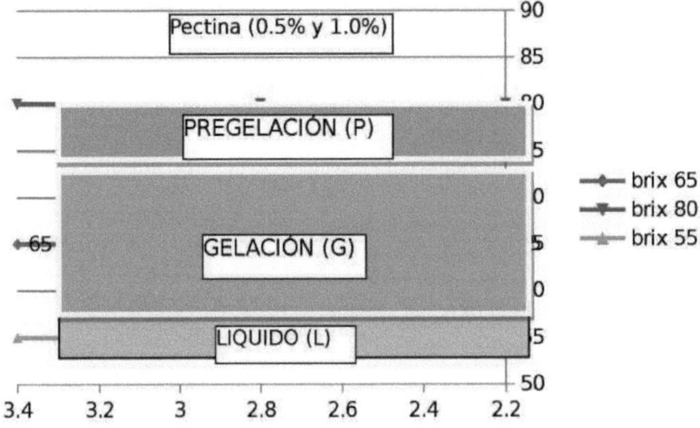

Figure 12. Range analysis for gels formed with pectin percentages (0.5-1.0%).

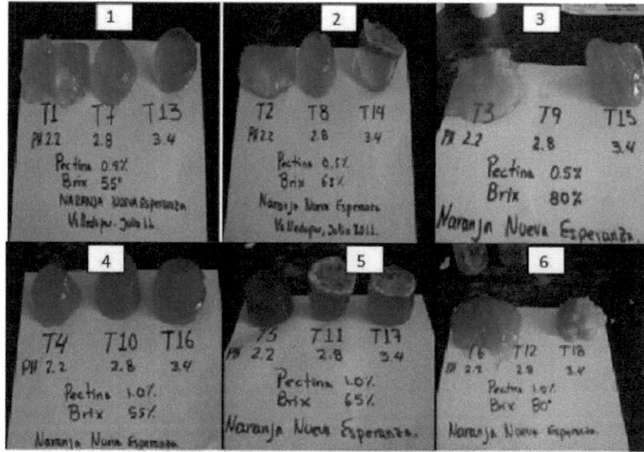

Photographic record of the gelation of all treatments.

Observing image 12, boxes 1, 2 and 3 show the gelation behavior at different pH, 0.5% pectin and the soluble solids content varies from 55 to 80^0 Brix. In box 1, it is observed that at pH 2.2 and 55 °Brix, there is gel formation, but it is a very weak gel, it is not consistent, its structure is homogeneous; at pH 2.8 and 3.4, the gel is stronger and soft consistency, but there is a slight syneresis (presence of water droplets). In box 2, at pH 2.2 and 55 oBrix, the behavior is similar to the previous one, i.e. a very weak gel, not consistent but homogeneous in its formation; at pH 2.8 and 3.4, the gel is stronger and more consistent and its structure is homogeneous and does not present syneresis, but at pH 3.4, a top layer of sugar crystals is present. In this case, the best gelling behavior occurs at pH 2.8, 65°brix (Figure 13). In box 3, at pH 2.2 and 80 oBrix, there is no gelation, but it is considered in a state of pregelation, the structure is non-homogeneous and of soft and viscous texture (image 14); at pH 2.8 and 3.4 pregelation is also present, with a better consistency than the previous one but soft, non-homogeneous and with a hard upper layer (image 15).

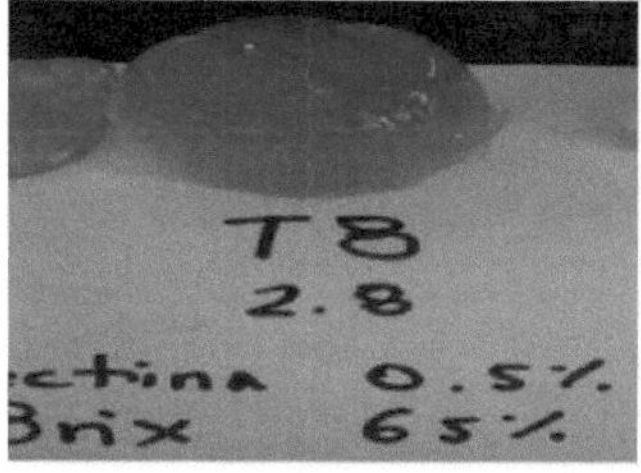

Gelation at pH 2.8, 65 °Brix and 0.5% pectin.

Pregelation at pH 2.2, 80 °Brix and 0.5% pectin.

Pregelation at pH 3.4, 80 °Brix and 0.5 % pectin.

Boxes 4, 5 and 6 show the gelation behavior at different pH, 1.0% pectin and varying soluble solids content from 55 to 80 °Brix. In box 4, it is observed that at pH 2.2 and 55 °Brix, there is gel formation, but it is a soft consistency gel and its structure is homogeneous; at pH 2.8 and 3.4, the gel is stronger, of moderately hard consistency, but there is a slight syneresis (presence of water droplets). In box 5, at pH 2.2 and 55 °Brix, the behavior is of gel, with soft consistency, homogeneous structure; with pH 2.8 and 3.4, the gel is stronger, moderately hard and hard consistency, homogeneous structure and no syneresis and the coloration is darker than when compared with 0.5% pectin, but with both pH, there is a superior layer of sugar crystals (image 16).

Image 16. Gelation and presence of sugar crystal layers at pH 2.8 and 3.4, 65^0 Brix and 1.0% pectin.

In this case, the best gelling behavior occurs at pH 2.2, 65 or Brix (Figure 17).

Gelation at pH 2.2, 65 oBrix and 1.0% pectin.

In Table 37, at pH 2.2 and 80 oBrix, there is no gelation, but it is considered to be in a state of pregelation, the structure is not homogeneous and the texture is soft and viscous; at pH 2.8 and 3.4 pregelation is also present, with a better consistency than the previous one but soft, not homogeneous and with hard layers (image 18).

Pregelling at 80 °Brix, 1.0% pectin and different pH

Table 37. Sensory evaluation of the gelling characteristics of the different pectin treatments. obtained from Valencian orange peel.

Treatment	CONDITIONS			Features		
	pH	^{0}BRIX	Pectin, %, %, %, Pectin	Consistency	Structure	Range
T1	2.2	55	0.5	1	homogeneous	Very soft gel
T2	2.2	65	0.5	1	homogeneous	Gel

T3	2.2	80	0.5	2	Viscous, non-homogeneous	pre gel
T4	2.2	55	1.0	2	homogeneous	Soft gel
T5	2.2	65	1.0	2	homogeneous	Gel
T6	2.2	80	1.0	3	Viscous, non-homogeneous	Pre gel
T7	2.8	55	0.5	2	Homogeneous (minimum syneresis)	Gel
T8	2.8	65	0.5	2	homogeneous	Gel
T9	2.8	80	0.5	2	Non-homogeneous (higher hardness)	Pre gel
T10	2.8	55	1.0	3	Homogeneous (minimum syneresis)	Gel
T11	2.8	65	1.0	3-4	Non-homogeneous (higher sugar hardness)	Gel
T12	2.8	80	1.0	3	Non-homogeneous (higher hardness)	Pre gel
T13	3.4	55	0.5	2	Homogeneous-less consistency than T7- very little syneresis	Gel
T14	3.4	65	0.5	2	Homogeneous - start of hard top sugar layer	Gel
T15	3.4	80	0.5	2	Non-homogeneous	Pre gel
T16	3.4	55	1.0	3	Homogeneous minimal syneresis	Gel
T17	3.4	65	1.0	4	Homogeneous - Start of hard coating	Gel
T18	3.4	80	1.0	3	top Non-homogeneous - top hard layer	Pre-gel

TEST OF THE DEGREE OF PECTIN GELATION

REPETICION	PH	ml solution	BRIX	gr Sugar	%PECTINA	gr Pectin	°SAG
RI	3	100	65	106,788 1	0,0486	0,1005	1062,568
R2	3	100	65	106,773 5	0,0970	0,2007	532,005
R3	3	100	65	106,782 6	0,1451	0,3004	355,468
R4	3	100	65	106,795 2	0,1931	0,4001	266,921
R5	3	100	65	106,780 0	0,2415	0,5006	213,304
R6	3	100	65	106,779 5	0,2896	0,6005	177,818
R7	3	100	65	106,780 1	0,3376	0,7004	152,456
R8	3	100	65	106,781 5	0,3854	0,8001	133,460
R9	3	100	65	106,780 0	0,4335	0,9002	118,618
RIO	3	100	65	106,781 0	0,4815	1,0005	106,728

Making the corresponding analysis of each experiment it was possible to determine that from repetition 6 onwards the gel remained at the bottom of the beaker when inverting its position, but it presented greater consistency from repetition 7 onwards, that is, the pectin analyzed presented a degree of gelation of 152°SAG, this data allows concluding that the pectin studied has a high gelling power; however, comparing these results with the values found by Rouse (1964) in the Valencia and Pineapple varieties that were 205 and 185 degrees USA.SAG respectively, we can say that the Valencia orange studied has a low value, but even so, we identify it as having a high gelling power.

6.2 TIME AND SPEED OF GELATION.

The results of the gelation times of the extracted pectin are shown below (Table 39), using pectin from treatment 1 (T1) from the Los deseos farm.

Table 39. Gelation time of the extracted pectin.

GELATION TIME OF THE EXTRACTED PECTIN	
REPEAT N	**TIME**
R1	14 min 45 sec
R2	16 min 13 sec
R3	18 min 32 sec
PROMEDIO	**16 min 30 sec**

From this we can conclude that the pectin analyzed is categorized as Medium rapid set pectin (Navarro and Navarro, 1985).

6.3 ESSENTIAL OIL EXTRACTION

6.3.1. Essential oil yield.

Table 40 shows the production of essential oils from the Los Deseos farm.

Table 40. Production of essential oil sample batch desires at different powers and 30 min of extraction - Chimichagua. 2009-2010

Id.	Watts	Water withdrawn, %	Performance of Dry base oil
T1	200	0	0
T2	400	0,22	0,07±0,007
T3	500	0,41	0,11±0,008
T4	600	0,62	0,58±0,05
T5	720	0,64	0,68±0,06

Source: pectin research seedbed

Evaluating table 40, it is observed that at 200 w no water or essential oil is extracted and that the greatest amount of water extracted is found at 720 w and 30 min. However, at this value and higher, the oil turns a strong yellow or even black color and solid material is carried away. It was observed that in the treatment with 600 w, the equipment changes the temperature from 94°C to 100°C after 15 min and the distillate is yellow, which indicates that from that time other types of compounds begin to distill and contaminate the sample. With respect to 700 w, the equipment changes the temperature from 94°C to 100°C after 10 min, causing the same phenomenon as above. It should be noted that at these powers and times the material is completely calcined.

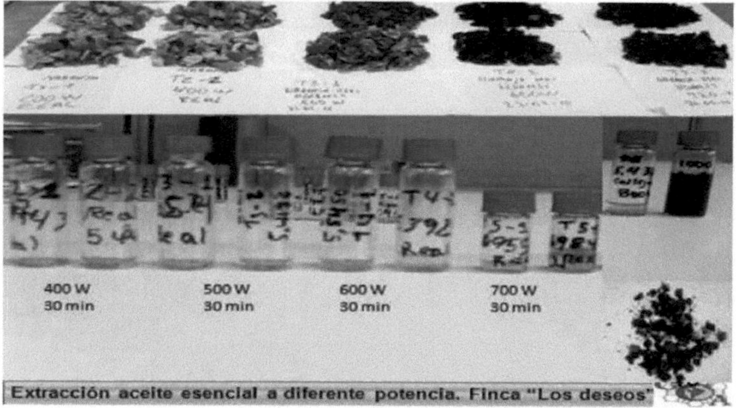

Extraction of essential oil of orange desires at different potencies.

Due to the above, the exposure time of the best treatment was reduced, obtaining yields similar to those of 500 w and 30 min, which means that at 600 w and an exposure time of 10 min, an oil dry yield of 0.58% is obtained, the boiling value of water in Medellín is not exceeded and the material remains in good condition and with little or no color change.

Tests were carried out under the same conditions as before, adding water (150 ml of water), and it was found that the amount of water added was not enough to distill and that the yield is very similar to that obtained without water, the sample remains the same color, with good texture but wetter than at the beginning and it is necessary to keep it so that there is no decomposition, which indicates that it is better to do it without solvent, since the costs of adding water are reduced and that from the engineering point of view it would not seem convenient to add water to have to distill it again.

6.3.2. Analysis of the essential oil extracted from the two batches.

The results of the majority components of the essential oils extracted from the two farms are shown below.

Table 41. Components of the essential oil extracted without solvent - "Los deseos" (30 min).

compound	400w(%)	500 w (%)	600 w (%)	720w (%)	600 w+ 150 ml (%)
Limonene	20 - 50,05	46,6 ± 82,8	95,6-98,5	52,3-53,8	19,74-66,25
Sabinene	7-18,1		0,12-0,21		34,28-35,14
beta.-Myrcene			0,23-1,59	0,92-1.04	
alpha - Phellandrene			0,04-0,05		
1R-.alpha.- Pinene			0,17-0,35		

Table 41 reports the most important components detected through gas chromatography coupled to mass, where it is highlighted that the limonene content is higher than 95.6 at 600 w and contrasts with the values lower than 66.2% of the other treatments. However, the opposite is true for sabinene, with the lowest values reported at 600 w.

Once the value of 600w was established as the adequate power, an experiment was carried out to determine the adequate exposure time to produce the greatest amount of oil and to prevent the shells from deteriorating. Figure 13 shows the behavior at different extraction times, whose values are similar up to 20 minutes and differ strongly at 30 min. However, as shown in Figure 14, after 10 minutes without solvent, the material begins to carbonize and at 30 min, it is completely carbonized.

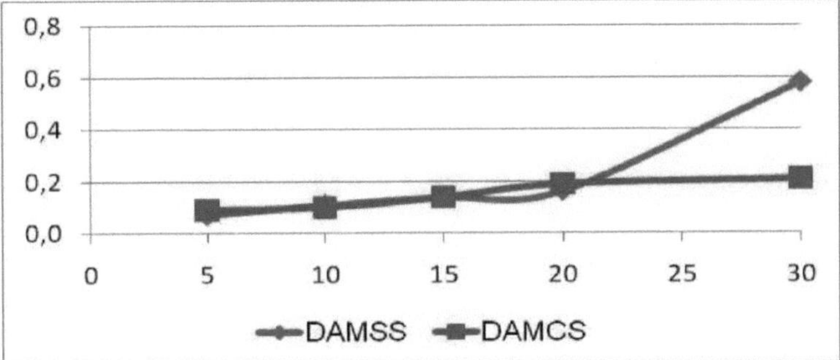

Figure 13. Oil yield at different extraction times with and without solvent - 600 W - "Los deseos".

It is therefore definitive that the best conditions for extraction are 600 w and 10 min of extraction, which is corroborated by Table 42, which shows the concentration of the components of the essential oil, especially limonene whose values are between 90.5-97.9 % at 10 min.

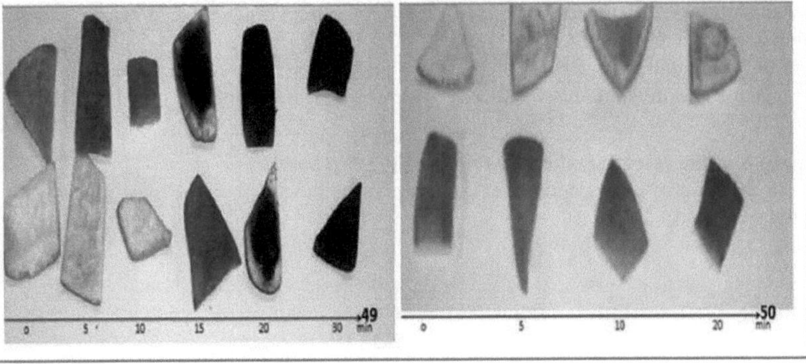

Figure 14. States of oranges at different times of essential oil extraction with and without solvent.

Table 42. Components of the oil extracted without solvent different times - 600W-"Los deseos".

compound	5 min (%)	10min(%)	15 min (%)	20 min (%)	30 min (%)
Oxalicacid , cyclobutyl hexadecyl ester	50,13-99,4	0	77,4-86,2	57,5-67,9	0
Alpha pinene	0	0,3-0,31	0,03-0,04	0,06-0,07	0,17-0,35
Limonene	0	90,5-97,9	11,8-18,5	32-33,6	95,6-98,5
Cyclotrisiloxane, 2,4,6-trimethyl-2,4,6-triphenyl	0,147				
Bacchotricuneatin	1,98				

Table 43 shows very low values of less than 66% limonene obtained from peels using water as solvent and microwave heating, which indicates that extraction without solvent is better and that from now on the tests will be carried out at 600w - 10 min with microwaves without solvent using water in situ in the peel as solvent.

Table 43. Components of the essential oil extracted with solvent at different times - 600 W- "Los deseos".

Beta Compound	5min (%)	10 min (%)	20 min (%)	30 min (%)
terpinene	0.08-0,09	0,24	6,0	14,28-24,14
Limonene	40.6 - 54.7	29,5	26,7-45.1	19,74-66,25
Beta myrcene			0,31	

In the "Los deseos" farm, the behavior of the extracted oil content is decreasing, with a higher content at maturity level 1 from 0.14% to 0.08% at maturity level 3 (see Figure 15), similar to that obtained by Reig (Reig Feliu A. 1943) and Di Giacommo (Di Giacomo A., Bovalo F and Postrino E. 1971).

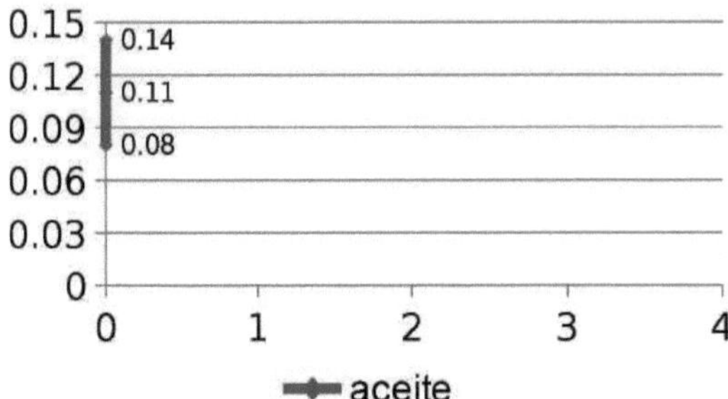

Figure 15. Behavior of essential oil at different stages of maturity, orange peels, Los Deseos.

Figure 16. Behavior of essential oil at different stages of maturity, orange peels, new hope.

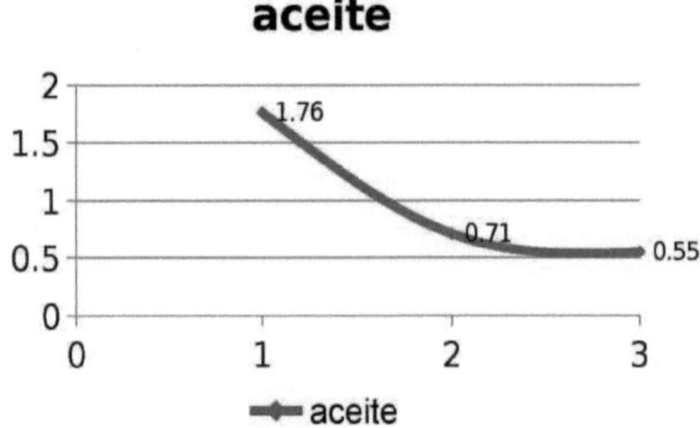

Figure 16 shows that the maximum oil content in dry basis of orange peel from the "Nueva Esperanza" soil is obtained at maturity stage one (1.76%), which decreases to 0.55%, these contents are even higher than those reported by other authors (yields close to 0.4%) (Ferhat, M; 2006; Rojas, J; 2009).

Table 44 shows a large variation in limonene content, possibly due to the non-homogeneity of soil characteristics, especially between the upper and lower part of the lot. The myrcene content increases at maturity levels 2 and 3, different from that obtained by some authors who reported that it did not change with maturity stage. For the "Nueva Esperanza" soil, Table 45, the stability in limonene content (94 - 97.6 %) may be due to the high homogeneity of the soil; myrcene content decreases with increasing maturity; sabinene, carene and a-phellandrene do not vary with maturity and a-pinene increases from maturity one to maturity two and remains stable.

Table 44. Compounds present in the "Los Deseos" lot at different degrees of maturity.

Compound	Maturity one	Maturity two	Maturity
Limonene	96.01-97.49	90.5 - 97.9	49.3-97.41
P-Myrcene	0.23-0.3	0.13-1.22	0.33-0.95
Sabineno	0.13-0.24	0.17-0.2	0.09-0.2
a-pinene	0.07-0.21	n.a.	0.24-0.31
a-felandreno	0.22	n.a.	n.a.
a-Tujeno	n.a.	0.20-0.21	0.086-0.09
Biciclo[3.1.1]hept-2-eno, 3,6,6-trimethyl	n.a.	n.d	0.1-0.28

n.a. not detected

As the agroecological conditions and genetic origin of the two orange lots are the same, the difference in orange quality and essential oil content of the "Nueva Esperanza" and "Los Deseos" lots can only be due to differences in the soil where the crop is established and management conditions.

Table 45. Compounds present in the "Nueva Esperanza" lot at different degrees of maturity.

Compound	Maturity one	Maturity two	Maturity three
Limonene	94.3-97.6	97.04-97.6	97.4-97.6
P-myrcene	1.66-1.92	0.93-1.15	0.9-0.96
Sabineno	0.17-0.25	0.14-0.18	0.17-0.18
1R a-Pinene	0.18-0.36	0.32-0.35	0.32-0.33
a-Felandreno	0.05-0.06	0.04-0.05	0.04-0.05
Careno	0.11-0.18	0.09-0.12	0.11-0.12

16.4. APPLICATION OF ESSENTIAL OILS AND PECTIN IN THE FOOD INDUSTRY ACCORDING TO THE RESULTS OF THE EVALUATION.

16.4.1. Pectin application.

According to the results we observed that the extracted pectin is classified as high pectin of medium and fast gelation since the results of the degree of esterification found were from 67.657 to 71.227% in the farm.

Nueva Esperanza (see Table 22) and from 62.867 to 77.910% at the Los Deseos farm (see Table 24), results that are within the ranges of 66 to 70% for pectin medium rapid set and 71 to 74% for pectin rapid set (see Table 4); these types of pectins are used for the manufacture of jams intended to be packed in small containers (maximum 1 kg), since the speed of gelation prevents the fruit pieces from floating during the cooling phase. These pectins are also used for those products that require a relatively high pH value (pH=3.0-3.5 for 65% soluble solids), as we found in our analysis that our pectin gels perfectly at pH 3.4 with 65% soluble solids (see table 37 and image 12).

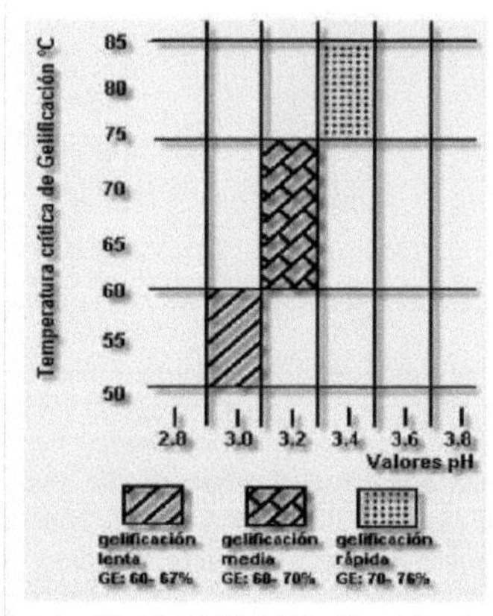

Figure 17. Temperature and pH ranges at which high methoxyl pectins gel.

It has also been successful, in the case of jams, a mixture of fast and slow gelling pectins to cause a gel to block at high temperatures the suspended fruit particles and also to allow final gelling at lower temperatures.

Figure 17. Presents the temperature and pH ranges at which high methoxyl pectins gel but of different gelling speed.

The dosage of pectin is easily calculated, theoretically, knowing its graduation or SAG degrees and the sugar content of the mass to be gelled: the ratio between the total weight of the sugars and the graduation of the pectin allows to obtain the quantity of pectin necessary for gelling.
In practice this dosage, valid for a syrup with 65°Bx and for a given pH, changes with the variation of its pH and the value of soluble solids.

The amount of pectin required to obtain a gel of a certain consistency is inversely related to the sugar concentration of the mass to be gelled as shown in Figure 18.

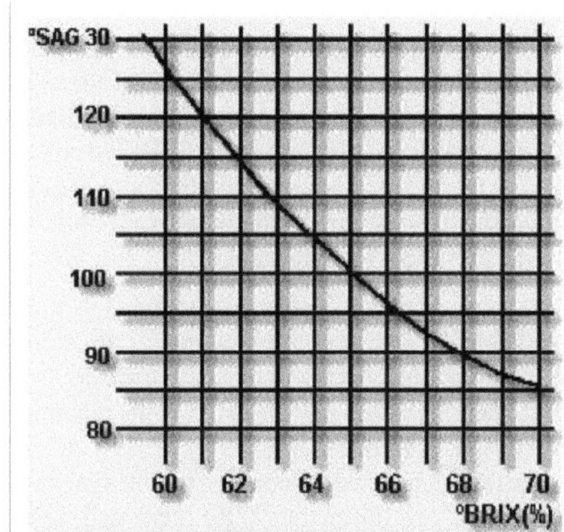

Figure 18 Equilibrium curve of gel consistency with varying pectin strength and final soluble solids of the product.

The greater the amount of sugar present, the lesser the amount of liquids, that is, the lower the density of the structure to retain it (and therefore less pectin), and vice versa, a lower sugar concentration requires a denser reticular structure (that is, more pectin) to retain the greater amount of liquids present; in our research we can corroborate this relationship by observing in image 12 boxes 1 and 4 where the Brix^0 of 55% is kept constant but the pectin concentration is varied from 0.5 to 1%, showing a greater consistency of the gel in box 4,

It was found that for low soluble solids, a higher amount of pectin is required to obtain adequate gelation. The same is true for boxes 2 and 5 for a constant brix of 65%.

Considering the behavior of the optimum pH of gelation with respect to the sugar concentration (Figure 19) the interdependence of the three components sugar-acid-pectin can be represented as in Figure 20.

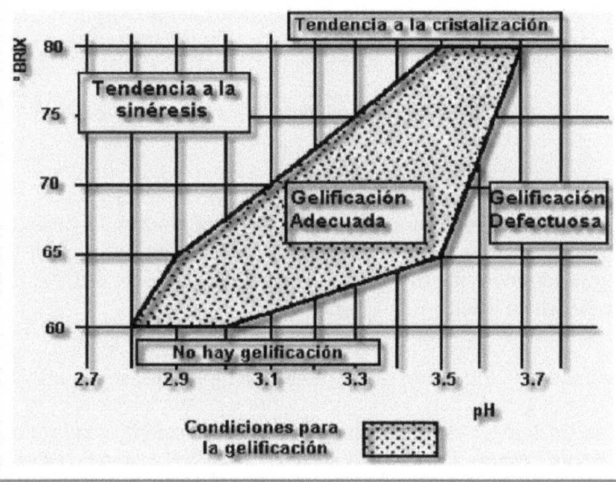

Figure 19. Gelation conditions of high methoxyl pectins.
Figure 19 shows an adequate gelling area between pH ranges and^0 Brix, this area coincides with the results obtained for the gelling power of the pectin evaluated in this research (see Table 37), therefore we can assure that our pectin will have an adequate gelling within the area shown in Figure 19, and would have its application for all types of food that keeps these ranges of pH and Brix degrees. In industrial practice, other factors intervene to modify the theoretical doses of pectin; these are due to the fruit, due to the gelling power of the natural pectic substances; due to the presence of soluble salts and insoluble fibers, which contribute to the consistency of the final product.

Figure 19 shows the inner area of the polygon, in which conditions of dry matter or solids concentration of the jam and pH are given in which gelation is more likely to occur. For example at 65° Brixx gelling can occur if the ingredient mixture fluctuates between pH 2.9 to 3.5. This pH range is significantly restricted if the Brix drops to around 60% or rises to 80%.
If a product of 68°Brix has a pH lower than 3.0 or higher than 3.6, it will probably present syneresis in the first case or defective gelling in the second. If the Brix are lower than 60% there will be no gelation and higher than 80% there will surely be crystallization of the sugar present in higher concentration, this is corroborated in the results found in table 37 and/or in image 12.

Figure 20 summarizes the interdependence of the three parameters, pectin, pH and Brix.
It is observed that mixtures with high Brix?? will gel more easily at pH 3.2 without needing high SAG pectins^0 and conversely, low Brix?? mixtures need more acidic pH (pH close to 2.8) with high SAG pectin or in general high amounts of pectin.

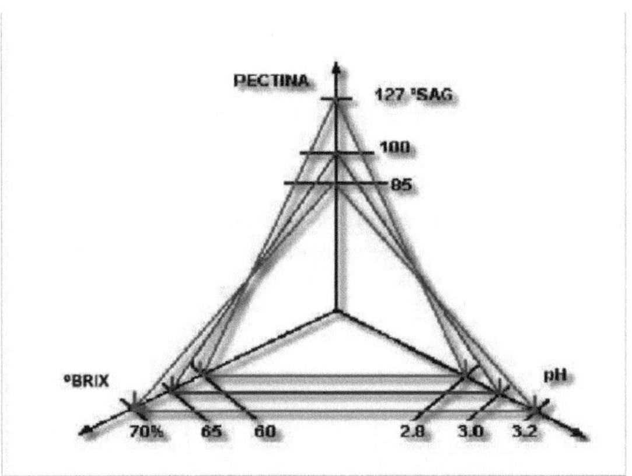

Figure 20. Balance of ingredients to achieve gelation.

Considering the difficulty of evaluating all the factors that modify the theoretical values, the exact dosage for each batch of fruit or juice is obtained by carrying out a small test, starting from the theoretical dosage and modifying it on the basis of the results obtained. Now, the optimum dosage will be valid for the whole batch.

A final factor, unrelated to the nature of the product components, which influences pectin dosage is the size of the packaging containers. Large jars require a greater consistency of the product than small containers, and pectin dosages vary accordingly. Thus, for example, 1 kg. containers would require a 2% increase in pectin dosage. They would need to increase the amount of pectin by 2%. A 10 kg.

container will be increased by 20%.

In the gelling process, the formation of the reticular structure of the gel takes place during the cooling phase that follows the cooking of the mixture of the various ingredients, and more precisely begins when the critical gelling temperature of the pectin used is reached. In practice, the theoretical values of this temperature are exceeded by a few degrees in the presence of natural fruit salts.
The temperature at which gelation occurs is higher if any of the following factors are increased: acidity, Brix, amount of glucose or pectin, and more, if it is high methoxyl and fast gelling (Figure 17).

Prolonged cooking causes, in addition to excessive inversion and caramelization of sucrose, a more serious drawback for pectin, which is its degradation and irreparable damage. Keeping the dough at temperatures above 100° C quickly affects the gelling qualities of pectin by producing its hydrolysis.

It is therefore very important, in order to use all the gelling power of pectin, to reduce to a minimum the time during which pectin participates in cooking and to accelerate the cooling of the finished product.

According to the conditions in which the pectin extracted and analyzed in this research should be handled, we can assure that this pectin can be used in the food industry in combination with sugars as a thickening or gelling agent, due to its high property of forming gels in an acid medium and in the presence of sugars. When pectin is heated together with sugar, it forms a network, which hardens upon cooling (Food-info, 2006).

It is generally used in the production of jellies, fruit preparations, fruit juice concentrates, fruit juices, desserts, ice cream, jams and fermented dairy products (Canteri, 2005).

In some products its use may be limited only to improve its consistency, but its potential use is to efficiently gel the products that keep the pH and Brix conditions previously established for this pectin.

6.4.2. APPLICATION OF ESSENTIAL OILS

These oils provide the food industry with characteristic flavors and aromas, widely used in bakeries, jams, confectionery, candies, soft drinks, ice cream, preservatives, cookies, dairy products, etc.

In addition to their use in the food industry, the use of essential oils in other industries, such as the chemical industry, the cosmetics industry and also in agro-industry as insecticides, bactericides and antifungal agents, has been booming.

It has been found that the antimicrobial action of essential oils by contact is efficient at high concentrations and for short times. Vapor levels between 0.1 to 0.9 mg/L in air could inhibit the growth of pathogenic bacteria (Inouye *et al*., 2001). Studies on the effectiveness of thyme and oregano oil vapors have shown that they are good bacteriostatic as well as bactericidal agents against *E. coli*, *S. typhimurium* and *P. aureginosa* (Pérez, 2006); as well as the vapors of cinnamon, garlic, lemon and orange oils are able to control the growth of molds such as *Aspergillus niger, Aspergillus parasiticus, Penicillium digitatum and Penicillium italicum* (Coronel, 2004).

The antimicrobial activity of citrus essential oils has been previously evaluated against bacteria and molds. O'Bryan *et al.* (2007) determined, by agar dilution, the minimum inhibitory concentration (MIC) of different orange essential oils against 11 strains of *Salmonella*, obtaining satisfactory results and MICs between 0.125 and 1%.

On the other hand, Viuda-Martos *et al.* (2008), by agar dilution, evaluated the antifungal action of lemon, mandarin, grapefruit and orange essential oils on molds commonly associated with food spoilage: *Aspergillus niger, Aspergillus flavus, Penicillium chrysogenum and Penicillium verrucosum*; concluding that the four essential oils studied show ability to reduce or inhibit the growth of the mentioned molds.

The capacity of these essential oils as inhibitors of bacteria and microorganisms is also highlighted. Therefore, in recent years there have been many investigations, which have demonstrated the antimicrobial power of essential oils, especially those extracted from citrus fruits; in these studies we can mention that of Dabbah *et al.* (Dabbah, 1970), who found that essential oils of mandarin, orange and grapefruit were shown to have antibacterial activity against bacterial strains of: Staphylococcus aureus, Escherichia coli, Bacillus subtilis, Pseudomonas aeruginosa and Salmonella among others.
These results make the study of essential oils relevant, especially citrus oils, due to their great importance for the pharmaceutical and food industry. It is known that essential oils, used as food additives, have antimicrobial effects and at the same time act as flavorings (Morales, 1996).

French (1985) points out that it is extremely difficult to correlate antimicrobial activity to a single compound or class of compounds, as the various components of any oil may act synergistically. Therefore, a comprehensive approach is necessary to explain the antimicrobial capabilities of an essential oil, whose performance could be the result of some quantitative balance of various components, where synergistic effects prevail (Caccioni *etal.*, 1998).

In the February issue of Trends in Food Science and Technology (Katie Fisher, 2008), an article is published in which the authors propose the use of citrus essential oils already used in food, specifically orange, lemon, lime and grapefruit, which have been shown to inhibit the growth of both gram (+) and gram (-) bacteria, in addition to being safe.

There are of course questions about the use of these substances, such as the effect of these compounds on the normal intestinal microbial flora. However, since the citrus oils mentioned above are already used in foods as flavorings, it is assumed that they should not produce an imbalance in the microflora.

What could constitute a limitation, or at least condition its use in any type of food, are its organoleptic characteristics.

The mechanism of action of citrus essential oils is still unknown. Some theories would indicate that they could induce morphological changes at the cell membrane level, which would be supported by some studies that demonstrated the disintegration of the outer cell membrane; others observed a thickening and dysfunction in the cell wall (Katie Fisher, 2008).
So far, the use of these essential oils in some foods, such as fish, meat and dairy products, has been investigated.

Fish: Application of citral and linalool reduced microflora in skin, intestines and gills of fish (carp) stored at 20 °C for 48 hours.

Meat: Lemon and orange essences in powder form were applied to meatballs at a concentration of 5%, demonstrating their effectiveness against the most frequent microorganisms in meat for 12 days.

Dairy: The tasting panels indicated that only lemon, orange and grapefruit essences would be acceptable for use in milk. However, only terpineol was found to be effective against Salmonella senftenberg, E. coli, S. aureus and Pseudomonas spp.

Among the major components of citrus essential oils, especially orange oil, the presence of more than 90% of the component d-limonene stands out; some of its possible applications are shown below.

Potential use of d-limonene as a replacement solvent for hazardous organic solvents and as an intermediate in transformations to useful products.
Limonene is used in the production of flavorings, flavoring agents, liqueurs, perfumes, toiletries, and as a raw material for pharmaceutical products and in organic synthesis. A variety of products or intermediates of other transformations can be synthesized from limonene, one of the most important being the transformation to para-cymene, which has applications in the fragrance, polymer and pharmaceutical industries and is also a solvent (www.chemsoc.org/pdf/gcn/limonenepractical.pdf).

Limonene is of interest to us because it is a naturally occurring product that has several applications and is presented as an environmentally friendly solvent, thus appearing as a solution to the use of volatile organic solvents, all of which are dangerous and polluting and have been widely used in industries and laboratories up to now.

Limonene has various applications:

a) . Replacement of toxic organic solvents.
Limonene can replace toxic, carcinogenic and highly flammable solvents in paints, lacquer removers, varnishes, rubber and plastic glues, such as benzene, toluene, xylene and chlorinated solvents. These solvents can also be replaced by limonene in laboratory practices such as recrystallizations, reactions and in all known applications of this type of solvents. In pathology laboratories limonene is a good alternative to replace xylene, since it is easily absorbed, metabolized and eliminated by the human body.

b) . Other industrial applications:
Among the important industrial applications, which have grown extraordinarily in the past decade, limonene, despite being used as a fragrance in the food and cosmetics industry, is used as a solvent in cleaning products of all kinds, both industrial and domestic, since mixed with water and a surfactant it produces a cleaner that is biodegradable, non-toxic and with a pleasant lemon scent, which is a great remover of tar and asphalt, since it can easily suspend them in order to be removed. For this reason, it is widely used for cleaning machinery, electronic parts, injectors, etc.
Since it is non-toxic and does not constitute a health hazard and its only known contraindication is the irritation it produces in the eyes, it is also used as a solvent in insecticides and in anti-parasite shampoos for dogs.

c) . Transformation to other substances useful in the industry
Other products or intermediates can be synthesized from limonene, one of the most relevant being the transformation to *para-cymene*, which has applications in the fragrance, polymer and pharmaceutical industries. Limonene is dehydrogenated to produce *para-cymene*, which can then be transformed to *para-cresol*, which is used in disinfectants, perfumes, preservatives or herbicides. These compounds also find application in the textile industry as cleaning agents.

CONCLUSIONS

J The analyses carried out on Valencia oranges grown in the department of Cesar, in the municipality of Chimichagua, yielded results in juice content higher than 50% from maturity stage two onwards, soluble solids above 10.5, acidity between 0.5 and 0.7 and a maturity index higher than 15. Ten percent of the samples analyzed have a juice percentage lower than 40%, although they have a C size and diameter greater than 63 mm, exceeding all the values established by standard 4086, which leads to the conclusion that from color two onwards all oranges meet the specific requirements demanded by the standard and can be consumed fresh, therefore this orange meets the requirements to be used in the industry.

J Under the context of this study, the best conditions for the extraction of essential oil from Valencia orange by microwave were 600 W and 10 min; the essential oil content was statistically equal with or without solvent. The essential oil content was lower in orange peel from the "Los Deseos" farm, with a yield of 0.11% at maturity stage 2, compared to 0.71% at the same maturity stage for the Nueva Esperanza farm.

J From the essential oil obtained by microwave radiation assisted hydrodistillation (HDMO), from the peel of sweet orange *Citrus sinensis osbeck* , valenciana variety, a species cultivated in the region of Chimichagua, Cesar, Colombia, the main volatile constituent was identified as the monoterpene known as limonene, with an average concentration of 95.76% at maturity stage 2.

J The essential oil extracted and analyzed in this study can be used in the food industry as a flavoring and to provide characteristic aromas widely used in bakeries, jams, candies, candies, soft drinks, ice cream, preservatives, cookies, dairy products; In addition to its use in other industries such as agribusiness as antifungal agents, it is concluded that the essential oil studied may have the ability to reduce or inhibit the growth of molds such as *Aspergillus niger, Aspergillus flavus, Penicillium chrysogenum and Penicillium verrucosum,* commonly associated with food spoilage.

J The pectic material content in the peel was 9.620 ± 0.764% (b.s.) for the fincalosdeseos farm and 9.968 ± 1.114 % (b.s.) for the nueva esperanza farm and is defined as the percentage of alcohol-insoluble colloidal substances composed mostly of polymers of galacturonic acid present in the peel.

J The pectin obtained from orange peel (*Citrus Sinensis Osbeck*) valencia variety from the Nueva Esperanza farm had a high degree of esterification ranging from 67.657 ± 0.870% to 71.227 ± 0.721% with an average of 69.442% and a high percentage of methoxyls of 13.004% and a purity of approximately 40.025% as galacturonic acid. According to these data, pectin will form gels with a high percentage of soluble solids and in acid medium.

J Due to its gelling speed, 16.5 minutes, the pectin obtained corresponds to the type commercially called "Medium rapid set" and within this classification it is a fast gelling pectin.

J Based on the property mentioned above, the pectin obtained may be used in the production of gels with suspended solids in which rapid gelation represents a uniform distribution of the product.

J The pectin obtained is suitable to be extended either with sugars and/or with inferior quality pectins in order to standardize the grade of the product to commercial levels of greater benefit to the producer.

RECOMMENDATIONS

- J Conduct a study concerning the optimization of the pectin extraction process to achieve higher yields, provided that the physicochemical and organoleptic characteristics of pectin are preserved, thus making it more viable to set up a plant for its extraction.

- J Conduct a study to determine the technical characteristics of limonene, the main constituent of the essential oil analyzed in this research to determine its technical and nutritional grade.

- J Conduct a feasibility study for the assembly and start-up of a plant for the extraction of essential oils and pectin from orange peel (*Citrus Sinensis Osbeck*) valencia variety, in the department of Cesar.

- J Given the results obtained regarding the quality characteristics of the orange analyzed in this research, which were optimal for commerce and industry, it is recommended that a feasibility study be carried out for the possible construction of a plant for the processing and transformation of this important raw material.

- J Conduct studies on yields and quality of pectin extracted from the same source at different stages of maturity, and in the two harvest periods where results are expected to vary.

- J To recover the ethanol generated in the precipitation stage and evaluate its reuse in the process.

- J Study various methods for the use of the waste generated, such as the bark of the husks and the bagasse from hydrolysis, to obtain compost material.

BIBLIOGRAPHIC REFERENCES

A. M. Bochek, N. M. Zabivalova, and G. A. Petropavlovski (2001). Determination of the Esterification Degree of Polygalacturonic Acid. Russian Journal of Applied Chemistry, Vol. 74, No. 5, 2001, pp. 796-799.Translated from Zhurnal Prikladnoi Khimii, Vol. 74, No. 5, 2001, pp. 775-777.
Adams, Robert (1995). Identification of essential oils components by gas chromatography/mass spectroscopy, Ed. Allured Publishing Corporation, Carol Stream, pp 469.
Adomako D. (1974), Chemical characterization of cocoa pectin. Chemical Industry; 21:873-874.
Al Di Cara (1983). Journal Essential Oils. In *Encyclopedia of Chemical Processing and Design*.
AOAC (1980). Official Methods of Analysis. Association of Official Analytical Chemists. Washington, D.C.
Barazarte, H., T. Garcia, L. Duran, L. Chaparro and J. Gámez. (2007). Evaluation and accuracy of m-hydroxyphenylphenol and carbazole methods applied in the quantification of pectic substances. Agrollania. 4: 53-62.
Barkman, T.J. (1997). Phytochemistry 44(5) 875.
Barrado, Andrés (1986). Analysis of food nutrients. Spain: Ed. Acribia.
Bhatia, B. S.; Khishnamurthy, G. V.; Lal, G. (1960). Preparation of Pectin from raw papaya (Carica Papaya) by aluminium chloride precipitation method; Food Technology; 13, 553.
Biocomercio sostenible (2003). Study of the Colombian Market for Essential Oils. Instituto de Investigaciones de Recursos Biológicos Alexander Von Humboldt, Bogotá, Colombia. 109p. www.humboldt.orq.co biocommerce.

Blanco Tirado, C. (1995). Journal Chromatography. 697A 501.
Blumenkrantz, N. and G. Asboe-Hansen (1973). New method for quantitative determination of uronic acids. Anal. Biochemistry. 5: 484-489.
Braverman, J. (1967). Introduction to the biochemistry of food. Barcelona, Spain: Ed. Omega. p. 111-124.
Burt, S. A. (2004). Essential Oils: Their antibacterial properties and potential applications in foods- a review. Int. J. of Food Microbiol. 94: 223. Cited in **Perez, T. F.** (2006). Effectiveness of thyme and oregano oil vapors as antibacterial agents. P.18. Master's thesis. Universidad de las Américas, Puebla, Mexico.
Cabeza M. (2004). Pectinase-producing microorganisms at low temperatures for winemaking, CONICET and Biotechnology Laboratory, Department of Biology - Food, Faculty of Sciences, University of San Martin, Argentina.
Cabra Rojas E (1988). Los Aceites Esenciales, Panorama Internacional y del Mercado Colombiano. *Tecnología*, **175** (5): 55-60.
Calvo, Miguel (2006). Pectins. Biochemistry of foods. [Page Internet]http://milksci.unizar.es/bioquimica/temas/azucares/pectinas.ht mljconsult: April 20, 2007].
Carbonell, E., E. Costell and L. Duran (1989). Evaluation ofvarious methods for measurement of pectin content in jam. J. Assoc. Off. Anal. Chem. 72(4): 689693.
Carbonell, E., E. Costell and L. Duran. 1990. Determination of pectin content in vegetable products. Rev. Agroquim. Tecnol. Tecnol. 30(1): 1-9.
Chaliha, B. P.; Barna, A. D.; Siddappa, G. S. (1963) Assam lemon as a Source of Pectin. Part I: Effect of method and pomace on the recovery and quality of pectin; Indian Food Packer, 17, Nos. 3, 1,
Copenhagen Pectin (1991). Molecular weight distribution of pectin. HPLC method. *Apl. Br.* **April**, 1428.
Coronel, C.P. (2004). Spice and seasoning extract vapors as antimicrobial agents. Master's thesis. Universidad de las Américas, Puebla. México.
Colombia International Corporation (2006). Permanent and annual crops by municipality. Cesar

2006.
Costa-Batllori. D. (2003). Natural antioxidants in animal feed. Inaugural lecture of the academic year 2003-2004. Royal Academy of Veterinary Sciences. Quoted in Ramírez, M. (2008). Vapor entrainment extraction and analysis of antioxidant properties of rosemary essential oil. Bachelor's thesis. Universidad de las Américas, Puebla, Mexico.
D'addonisio R., Marín M., Ruesga L., Vasquez R. (2008). Faculty of Engineering, University of Zulia, Venezuela. Pectin extraction from plantain (Mussa AAB, plantain subgroup) harton clone. Revista facultad agronómica, Vol. 25 pp. 318-333.
Dabbah, R., V.M. Edwards and W.A. Motas (1970). Antimicrobial Activity of Some Citrus Fruits Oils on Selected Food-Borne Bacteria. Appl. Microbiology. 19 (1): 27-31
Denayer, R., Tilquin, B. (1994). Riv. Ital. Eppos 5 7-12. CA 122(1995)247999x. Devia, Jorge (2003). Process to produce citric pectins. In: Revista Universidad Eafit. Universidad Eafit. Medellín: (Jan-Feb-Mar, 2003); p. 21-29 **Di Giacomo A.,Bovalo F and Postrino E**. (1971). Sull' essenza di arancia prodotta industrialmente dai frutti della piana di rosarno. Essenz. Der. Agrum. 41:239
Diaz J A (2002). Analysis of the international market for essential oils and vegetable oils. Alexander Von Humboldt Institute-Biocomercio Sostenible. Bogotá.
Dugo, G. (1992). Journal Essential Oil Resources, 4 589-594.
Dugo, G. (1992). Perfum. Flavor. 17 (5) 57-74.
National Agricultural Survey (ENA) 2009. CCI - MADR.
Eskin, N.A.M. (1971). Biochemistry of foods. United States: Ed. Academic Preus.
Espinal CF, Martínez HJ, Peña Y. (2005). The citrus chain in Colombia: A global view of its structure and dynamics 1991-2005. Bogotá: Ministry of Agriculture and Rural Development, Observatorio Agrocadenas Colombia. [Web site]. Available at: http://www.agrocadenas.gov.co. Accessed: 15 March 2005.
URPA Statistics - Secretary of Agriculture of the Department of Cesar. 1993.
Estrada, Amparo (1998). Citrus pectins. Effect of steam entrainment on extraction and different drying methods. In: Revista Noos. National University of Colombia. Manizales: (7, Dec., 1998); p. 23-34.
Study of the Colombian market for essential oils. Alexander Von Humboldt" Biological Resources Research Institute. 2003. Colombia. www.humboldt.orq.co. www.humboldt.orq.cobiocomercio.
FAOSTAT | © FAO Statistics Division. July 2009.
Ferhat, M.; Meklati, B.; Smadja, J.; Chemat F. (2006). An improved microwave Clevenger apparatus for distillation of essential oils from orange peel. *JournalofChromatography.* A, 1112, pp 121-126.
Fisher, K. and Phillips, C. (2008). Potential antimicrobial uses of essential oils food: is citrus the answer? Trends on Food Science and Technology. 19:156.
Fiszman S.M., Costell E., Duran L. (1984). The rheological behavior of hydrocolloid gels. Relationship to their composition and structure. Journal. Agroquim Food Technology. Vol. 24, No 2 p. 177-189 (1984).
Francis, B.J. and Bel, K. J. (1975). Commercial pectin; a review. In: Tropical Science: 17;p. 25-44.
Gaviria, N., López, L. (2005). Laboratory scale extraction of passion fruit pectin and preliminary scale-up to pilot plant. Graduate work. Eafit University. Medellín.
Genu Pectin Co. (1979). Kobenhavns Pektinfabrik. Technical manual of the Genu pectin factory in Denmark. pp 65.
Gierselmer K. Pectin and pectin enzymes in fruit (1997). Vegetables technology. 118, pp171 -185.
Giraldo V, Celina (1991). Obtaining pectin from tree tomato. Basic engineering. Degree thesis. National University. Faculty of Mines. Medellín.
Gómez Z., Juan F. (1998). Technical Feasibility of the Isolation and Characterization of Citric Pectin for the Agroindustrial Sector (Degree Work). Medellín: Corporación Universitaria Lasallista, School of Administration.

Griffiths, D. W. (1992), Phytochemistry Anal. 3 250-253 (1992).
Grosse R et al (2000). Extraction of the Essential Oil of Cajera citrus orange. *Acta Científica Venezolana* **51**(2), 200-208.
Günther, E. 81984). *The Essential Oils.* Vol. 1: History and origin in Plants Production Analysis. Krieger Publishing: New York, USA.
Guzmán, R., A. Suárez and C. Castro. Castro (1977). Determination of pectin content in mango and its application in jam production. University of Bogota Newsletter. Colombia.
Harborne, J.B., "Phytochemical Methods" (1973), Chapman & Hall, London, 92-105 pp.
Heath, H.B. and Reineccius, G. **(1986).** *FlavorChemistryand Technology*. AVI HERBSREITH. The apple. The pectin.
Herbstreith & Fox (2001). The Specialists for Pectins. http://www.herbstreith-fox.de/produkte/englisch/einstant.htm (10feb. 2001).
Hercules, Food Gums Division (sf). General description of pectin. (sl). (se.). p. 3-21.
Ibarz, A., A. Pagán, F. Tribaldo and J. Pagán. (2006). Improvement in the measurement of spectrophothometric data in the m-hydroxydiphenyl pectin determination methods. Food Control 17:890-893.
IFT Committee on Pectin standardization (1959). Final report. In Food Technology Journal. 13: 496-500.
Ingham, B.H. (1993). Journal Agriculture Food Chemical. 41 951-954.
Inouye, S., Takizawa, T. and Yamaguchi, H. (2001). Antibacterial activity of essential oils and their major constituents against respiratory tract pathogens by gaseous contact. J. Antimicrobial Chemotherapy. 47: 565-573.
Alexander von Humboldt **Biological Resources Research Institute.** Sustainable Biocommerce (2003). Study of the Colombian market for essential oils. Bogotá, p109. [Web site in Internet]. Available at http://www1.minambiente.gov.co/viceministerios/ambiente/mercados_verdes/IN F0%20SECT0RIAL/Mercado%20nacional%20de%20aceites %20esenciales.pdf.Accessed: October 10, 2008.
Journal Chemical Education. 68 267 (1991).
Jennifer P. Rojas Ll., Aidé Perea V., Elena E. Stashenko. Stashenko (2009). Obtaining essential oils and pectins from citrus juice by-products. Vitae, journal of the faculty of pharmaceutical chemistry. ISSN 0121-4004 Volume 16, 1, pp. 110-115.
Jennings, W. and Shibamoto, T. (1980). Qualitative analysis of flavor and fragrance volatiles by glass capillary gas chromatography, Ed. Academic Press, London, p. 472.
Jiménez, M.C., Soto, J. and Villaescusa, M.A. 2006. "Physical Chemistry for Chemical Engineers". Editorial de la UPV. Valencia. Valencia. Spain.
Jittra, S., N. Suwayd, S. Cui, and D. Goff. 2005. Extraction and physicochemical characterization of krueo Ma Noy pectin. Food hydrocolloids. 19(5):793-801.
Joseph. G. H. and Baier, W. E. (1949). Methods of determining the firmness and setting time of pectin test jellies. Food Technology Journal. 3:18-22.
Katie Fisher, Carol Philips (2008). *Potential antimicrobial uses of essential oils in food: is citrus the answer?* Trends in Food Science & Technology, **19**, 3, 156-164.
Kertesz, Z.I. (1951) "The Pectic Substances." Interscience. New York.
Kim WC, Lee DY, Lee CH, Kim CW (2004). Optimization of narirutin extraction during washing step of the pectin production from citrus peels. Journal Food Engineering; 63 (2): 191-197.
Kitner, P.K. and J.P. Van Buren (1982). Carbohydrate interference and its correction in pectin analysis using the m-hydroxydiphenyl method. Journal Food Science. 47: 756-759.
Konig, W.A. and Joulain, D. (1998).The atlas of spectral data of sesquiterpene hydrocarbons, Ed. Verlag, Hamburg, 658 p.
Kovats, E. Helv. Chim. Acta, 1958, Vol. 41, 1915p.
Liu Y, Shi J, Langrish TAG. (2006). Water-based extraction of pectin from flavedo and albedo of orange peels. Chemical Engineering Journal (3); 120: 203-209.
Lopez J B, Jean F, Gagnon H, Collin G, Garneau F, Pichette A (2005). *Journal Essential Oil*

Resources. **17**: 1-7.
Maat, L. (1992). Journal Essential Oil Resources. 4 615-621 (1992).
Marín RF, Soler C, Benavente O, Castillo J, Pérez JA (2007). By-products from different citrus processes as a source of customized functional fibers. Food Chemistry. 100(2): 736-741.
Martínez, A. (2003). Essential oils. *Revista Universidad de Antioquia*, pp. 1-34.
May, C.D. (1990). Industrial pectins: sources, production and applications. *Carbohydrate Polymers* **12**, 79-99.
McCready, R. and E. McComb (1952). Extraction and determination of total pectin materials in fruits. Anal. Chem. 24:1986-1988.
McCready, R. and E. McComb (1970). Pectin on methods in Food Analysis. Per Joslyn, M.A. New York, Academic Press. Chap. XIX pp. 565-597.
Merck Chemicals (2009). Limonene. Available: http://www.merk- chemicals.com.mx
Mesbahi G, Jamalian J, Farahnaky A. (2005). A comparative study on functional properties of beet and citrus pectins in food systems. Food Hydrocolloid. 19 (4): 731-738.
Michel F, Thibault J-F, Mercier C, Heitz F, Pouillaude F. Extraction and characterization of pectins from sugar beet pulp. Journal Food Science, 1985; 50: 1499-1502.
Miranda, M.E. (1993). Characteristics, production and use of pectins. In: Revista alimentación, equipos y tecnología, Year n^0 12, N^0 9, pp. 61-66.
Miyamoto A, Chang KC (1992). Extraction and physicochemical characterization of pectin from sunflower head residues. Journal Food Science; 57: 1439-1443.
Mondello, L. (1994). ChromatographiA, 39 529-538.
Mondello, L. (1996). Journal Chromatography Science. 174-181 (1996).
Monsoor MA, Proctor A. (2001)Preparation and functional properties of soy hull pectin. Journal American Oil Chemical Society, 78(7):709-713.
Morales de Godoy, V. (1996). Extraction and Characterization of the essential oil of Tahitian Lime Citrus aurantiifolia (Chritms) Swingle. Special Degree Work. LUZ. Experimental Faculty of Sciences, Maracaibo, Venezuela. pp. 18-19.

Morin, Ch. 1985. Citrus cultivation. 2 ed. San José, Costa Rica: IICA. 607p.
Navarro, G.; Navarro, S. (1985). "Pectic substances: chemistry and applications". Secretariat of publications and scientific exchange. University of Murcia.
O'Bryan, C. A., Crandall, P. G., Chavola, V. I. and Rickie, S. C. 2007. Orange essential oils antimicrobial activities against *Salmonella spp.* Journal Food Science, 73(6): 264-267.
Obipectin. Pectin. [Online], Switzerland, [cited Jan. 11, 2005]. Available from World Wide Web: http://www.obipektin.ch/eZprodukte/set pektin.htm.
Ochocka, J. R (1997),Phytochenistry, 44(5) 869,
Food and Agriculture Organization of the United Nations (1975) Standards of identity and purity of certain food additives, flavour enhancers, thickeners and others. Rome.
Ortiz, Claudia and Acevedo, Fernando (2002) Pectinase production by submerged fermentation using citrus wastes as substrate (Project financed by COLCIENCIAS, PMBTA and OAS). Medellín.
Ortuño, M.F. (2006). Practical Manual of Essential Oils, Aromas and Perfumes. Aiyana Ediciones. Spain.
Pagan J. (1998). Enzymatic degradation and physical and chemical characteristics of peach bagasse pectin, Biblioteca Virtual Miguel de Cervantes, **Pauli, A.** (2001). Antimicrobials properties of Essential oil constituents. International Journal Aromatherapy. 11(3): 126.
Pérez, T. F. (2006). Effectiveness of thyme and oregano oil vapors as antibacterial agents. Master's thesis. Universidad de las Américas, Puebla, Mexico.
Pilgrim, G.W. (1991). Jams, Jellies and Preserves. In: The Chemistry and Technology of Pectin. San Diego CA: Academic Press.
Pilnik, W. and Voragen, A. G. (1970). Pectic substances and other uronides. In Hulme, A.C. ed. The Biochemistry of Fruits and their products. London, Academic Press. Vol. 1. pp. 53-80.
Departmental Development Plan. A Cesar for all. 2008-2011.

Praloran, J. C. (1977). Los agrios. Barcelona, Spain: Blume p.520. Publishers, USA.
Pruthi, J. S.; Mookerji, K. K.; Lal, G. (1960). Studies on the Dehydration of Guava for Subsequent Recovery of Pectin during off-season; Central Food Tech. Res. Inst., India.
Puerta, Jimmy Andrés (2001). Obtaining pectins from the peel of apple (malus domestica Borkh) variety ANNA. Noos Magazine. National University of Colombia. Manizales. p. 137-147.
Quimerco S.A., 2004
Ramirez, Jairo (1980). Parameters for the extraction and characterization of tree tomato pectin. Degree Project. National University. Dept. of Chemical Processes. Medellin.
Reig Feliu A. (1943). Quantitative analysis of the fruit essences of the most cultivated citrus fruits in Spain. Bol. Inst. Nal. Invest. Agron. Notebook N0. 33
Rodríguez, M.D., A. Redondo and M.J. Villanueva. 1992. Comparative study of m-hydroxyphenylphenol and 3,5-dimethylphenol methods to determine pectic substances in turnip (*Brassica napus*). Alimentaria. 79: 79-82.
Rojas, J.; Perea, A.; Stashenko E. (2009). Obtaining essential oils and pectins from citrus juice by-products. *Vitae*. Vol16 No.01. 110115.
Ros, J.M., D. Saura, L. Coll and J. Laencina. 1992. Advanced analytical methods for the determination of pectic substances and pectolytic enzymatic activities. Rev. Alim. Equip. Tecnol. 2: 149-155.
Rouse and Knorr, L. C. (1970). Evaluation of pectins from Florida lemons harvestedfrom young trees. Prosc. Fl. State. Horticulture. Society. 83:281-284. **Rouse, A. H. Atkins, C. D. and Moore, E. L.** (1964). Evaluation of pectin in component parts of Pineapple oranges during maturation. Proc. Fl. State Hortic. Society. 77: 221-273.
Saldarriaga, Francisco (1974). Determination of pectins in citrus fruits: sweet orange (citrus sinensis), sour orange (citrus auramtimu), grapefruit (citrus paradisi) and lemon (citrus limón), grown in Cotove (Sta. Fe de Antioquia). Degree thesis. National University. Faculty of Agricultural Sciences. Medellín. **Sánchez R, Pino J, Chang L, Roncal E, Rogert E** (1994). De-terpenation of orange essential oil by extraction with dilute ethanol. *Alimentaria* **249**: 59-61.
Sánchez, F. J. (2006). Extraction of Essential Oils: Colombian Experience. II International Congress of Medicinal and Aromatic Plants: National University of Colombia, Palmira: 1-8.
Schreier, P. (1984), Analysis of volatiles, methods and applications, Ed. Walter de Gruyter, New York, **1984**, 1463 p.
Shrivas, S. R.; Pruthi, J. S.; Siddappa, G. S. (1963). Effect of Stage of maturity of Fruit and Storage Temperature on the Volatile oil and pectin content of Fresh limes; Food Science, 12, 340, 1963.
Sierra L., Bolaños S. (2006). Design of a process to obtain and purify pectin from coffee processing wastes, Universidad EAFIT, Colombia.
Stahl, E. ed., "Thin Layer Chromatography. A Laboratory Handbook", 2nd. edition, (1969). Springer-Verlag, Berlin-Heildelberg-NewYork, 236-239 pp.
Stashenko, E.E., Combariza, M.Y. and Puertas, M. A. (1988). Essential oils: Extraction and analysis techniques. Laboratory of Photochemistry & Chromatography, School of Chemistry, Faculty of Sciences, UIS, Bucaramanga, **1998**.
Stashenko, E.E., Martinez, R., Pinzón, M. E. and Ramirez, J. (1996). Changes in chemical composition of catalytically hydrogenated orange oil (citrus senensis). Journal ofChromatography A. 752(1-2): 217.
Ullman, F. 1950. Encyclopedia of Industrial Chemistry. Barcelona, Gustavo Gili, t. ll, 748 p.
Visciglio, S.B. and N. San Juan. 2000. Determination of pectin content in citrus fruit peel: modification and optimization of the carbazole method. S. Cs. Eng. Alim. 2: 719-733.
Viuda-Martos, M., Ruiz-Navajas, Y., Fernandez- Lopez, J. and Perez-Alvarez, J. (2008). Antifungal activity of lemon (citrus lemon L.), mandarin (citrus reticulata L.), grapefruit (citrus paradise L.) and orange (Citrus sinensis L.) essential oils. Food Control. 19(12): 1130-1138.
Vogler, B. (1998). Journal Natural Production, 61 (2), 175.
Wagner, H., Bladt, S., Zgainski, E. M. (1984). "Plant Drug Analysis", Translated into English by Th. A. Scott, Springer-Verlag, Berlin-heildelberg-New York-Tokyo, pp 5-8.

Wang S., Chen F., Wu J., Wang Z., Liao X., Hu X. (2005). Optimization of pectin extraction assisted by microwave from Apple pomace using response surface methodology. Journal offood engineering. Vol. 78 pp. 693-700.
Weiss E A (1997). Essential Oil Crops. Cab International: New York, USA, pp. 417-511.
Wilkins MR, Widmer WW, Grohmann, K, Cameron RG (2007). Hydrolysis of grapefruit peel waste cellulase and pectinase enzymes. Bioresource Technology. 98 (8): 1596-1601.
Willats, W.G., J.P. Knox and J.D. Mikkelsen. 2006. Pectin: new insights into an old polymer are starting to gel. Trends Food Science. Technology. 17: 97-104.
Yáñez L, Lugo D, Parada Parada, Y. (2007). Study of the essential oil from the peel of sweet orange (citrus sinensis, valencian variety) grown in labateca (norte de santander, Colombia). Bistua: Revista de la Facultad de Ciencias Básicas, vol. 5, 1. Universidad de Pamplona, Bucaramanga, Colombia. pp. 3-8.
Knoor, L. C. (1969). Maturity Changes in pectic substances and citric acid of Florida lemons. Prosc. Fl. State Horticulture. Soc. 82:208-212.
Yates, Grand Prairie (1999). Aloe pectins. U.S. Patent No. 5,929,051. July 27, 1999.
Yegres S., Sánches J., Belmar M., Riberos W. and Belmar D. (2001). Production of pectic enzymes preliminary trials. Saber, Vol. 13 pp. 5559.

ANNEX 1.

Physicochemical analysis of Los Deseos valence oranges

Physicochemical analysis of Los Deseos valence orange - Part 1											
PUNT 0	#	PES Og	e m m	CALIB RE	COLO R NTC	WEIGHT g CASCA RA	WEIGHT g SEMIL LA	WEIGHT g JUICE	PH JUG O	ACIDE Z JUICE %	°BR IX
P1	1	107, 4	61	E	1	56,690	3,73	41,32 1	3,82		10, 7
P1	2	142, 1	66	D	1	64,540	4,113	69,24 1	3,55		10, 3
P1	3	206, 1	74	C	1	94,641	3,771	103,6 37	3,75		10, 8
P1	4	163, 4	70, 5	D	2	80,105	5,162	72,55 6	3,78		11
P1	5	205, 2	74	C	1	75,78	3,331	121,0 25			
P2	1	225, 1	77	C	1	154,05 1	2,627	56,66 4	3,28		8,5
P2	2	153, 4	65	D	2	82,210	4,901	57,48 2	3,49		11, 5
P3	1	145, 3	61, 5	E	2	57,451	5,100	60,87	3,71		8,7
P3	2	126, 9	65	D	3						
P3	3	147, 9	66	D	3	49,869	0,549	81,79 7	3,38		10
P3	4	149, 4	67	D	0	87,019		25,90 5	3,08		6,5
P3	5	161, 8	70, 5	D	3	89,099	6,923	93,03 5	3,77		9,7
P3	6	201, 8	74, 5	C	1	82,884	5,513	106,1 08	3,79		9,4
P3	7	192, 7	73	C	1	92,140	6,363	88,04 3	3,61		11, 2
P3	8	277, 7	76	C	3	84,779	3,513	124,5 38	3,79		9,6
P3	9	199, 7	73	C	2				4.03	0,54	10, 0
P3	1 0	216	75	C	3	83,596	4,842	119,6 17	3,84		7,9
P3	1 1	227, 6	75	C	5	112,67 5	5,246	101,6 05	4,31		12, 7
P3	1 2	187, 3	77	C	5	148,23 6	1,32	31,01 7	4,04		12, 8
P3	1 3	181, 6	72, 5	C	4	91,541	0,604	76,81 6	4,12		12, 3
P3	1 4	189, 5	75	C	5						
P3	1 5	182	70	D	2	86,893	5,79	84,81 3	3,58		9,3
P3	1 6	244, 2	80	C	4			118,5 45			

P3	11	273, g	S3	C	A	11S,A92	6,112	1AA,333	3,6S		1G
P3	1S	rgб, 1	S3, 3	B	A				A,2S	G,SA	12, S
P3	1 g	r13, S	SS	B	A	11б,Aб G	1,3S1	S1,2S1			
P3	г0	2S3, 1	S2	C	A				A,32	G,S2	1A, G
P3	г1	го3, S	1S	C	2				A,AS	G,A1	12, A
P3	гг	22S, A	1S	C	1	g1,AS1	S,G3S	121,3 9A	3,S		9,2
P3	г3	198, A	12, 2	C	1				3,SS	G,16	1G, б
P3	гA	23O, 1	1S	C	A				A,SS	G,A1	16, б
P3	гS	19s, г	13, S	C	3				3,62	G,S2	g,s
P3	гб	г3б, A	SG	C	A				A,32	G,S	1A, G
P3	г1	гм, g	16	C	3						
PA	1	1AA, A	6S	D	G	1S,2S3	3,Aб9	б3,бб A	3,GA		9,1
PA	г	1S1, g	1G	D	1	Sб,og1	2,S29	9S,GG g	3,AA		g,S
PA	3	1AA, 3	61	D	1	1S,SgA	A,G2	61.1S S	3,6S		1G, 3
PA	A	г03, 1	13	C	2	gS,323	1,1Sg	9S,AS g	A,3S		12, S
PA	S	1SG, 1	6S	D	1	11^11	S,2SS	б9,бб S	3,AA		1G, S
PA	б	1S6, б	61	D	G	91,22s	3,11S	SS,1G б	2,91		g,s
PS	1	13A, g	6S	D	1	SG,26A	1,S1S	бA,б9 S	3,SA		
PS	г	111, 1	6S	D	1	61,S2g	S,S23	69,G11	3,1б		1G, б
PS	3	12A, 1	63, S	D	1						
PS	A	1G2, S	62, S	D		6S,16S	3,G9	AA,9S g	3,33	1,A	1G, б
PS	S	1SG, S	6S	D	1	66,ssg	1,1SS	1S,G1 б	3,SS		1G, S
PS	б	1AA, S	бб	D	1	S3,A11	2,A31	19,31 б	3,1A	G,1A	1G, 1
PS	1	12g, S	62	D	1	AA,13S	3,3GA	бA,3AG	3,SA	G,S1	1G, S
PS	S	11G, б	1G	D	1				3,1G	G,69	11, г
PS	g	1S6, g	б1, S	D	1				3,12	1,11	11, г
PS	10	1S1, г	6S, 3	D	2	6S,913	A,3SA	1S,gS g	A,1A		1G, 1
PS	11	16s, б	1G	D	1	6S,1S1	A,S1S	SS,26 g	2,S3	G,1A	11, G
PS	1г	1S6, g	6S, S	D	2	S1,S1A	A,A1g	6S,9A A	3,S		11, 3

<table>
<tr><td>P5</td><td>13</td><td>161, 4</td><td>69</td><td>D</td><td>2</td><td>68,274</td><td>3,759</td><td>84,90 0</td><td>3,76</td><td></td><td>10, 4</td></tr>
<tr><td>P5</td><td>14</td><td>214, 8</td><td>76</td><td>C</td><td>1</td><td></td><td></td><td></td><td></td><td></td><td></td></tr>
<tr><td>P5</td><td>15</td><td>259, 4</td><td>80</td><td>C</td><td>4</td><td></td><td></td><td></td><td></td><td></td><td></td></tr>
<tr><td>P5</td><td>16</td><td>206, 9</td><td>75</td><td>C</td><td>0</td><td>131,62 0</td><td>4,488</td><td>67,31 0</td><td>3,03</td><td></td><td></td></tr>
<tr><td>P5</td><td>17</td><td>178, 6</td><td>70</td><td>D</td><td>1</td><td>68,188</td><td>5,242</td><td>93,04 0</td><td>3,64</td><td>0,77</td><td>11, 0</td></tr>
<tr><td>P5</td><td>18</td><td>243, 9</td><td>82</td><td>C</td><td>3</td><td></td><td></td><td></td><td></td><td></td><td></td></tr>
<tr><td>P5</td><td>19</td><td>240, 1</td><td>78</td><td>C</td><td>4</td><td>96,170</td><td>4,138</td><td>120,287</td><td>3,36</td><td>0,5</td><td>12, 5</td></tr>
<tr><td>P5</td><td>20</td><td>274, 3</td><td>79, 5</td><td>C</td><td>4</td><td>114,048</td><td>6,124</td><td>130,401</td><td>3,42</td><td>0,5</td><td>12, 6</td></tr>
<tr><td colspan="12">Physicochemical analysis of Los Deseos Valencia orange - Part 2</td></tr>
<tr><th>PUNT 0</th><th>#</th><th>ESP 0g</th><th>e m m</th><th>CALIB RE</th><th>COLO R NTC</th><th>WEIGHT g CASCA RA</th><th>WEIGHT g SEMIL LA</th><th>WEIGHT g JUICE</th><th>PH JUG O</th><th>ACIDE Z JUICE %</th><th>°BR IX</th></tr>
<tr><td>P5</td><td>21</td><td>260, 2</td><td>80, 5</td><td>C</td><td>4</td><td>141,122</td><td>2,352</td><td>110,395</td><td>4,32</td><td></td><td>11, 2</td></tr>
<tr><td>P5</td><td>22</td><td>240, 7</td><td>77, 5</td><td>C</td><td>4</td><td>104,498</td><td>2,897</td><td>128,477</td><td>4,28</td><td></td><td>12, 3</td></tr>
<tr><td>P5</td><td>23</td><td>251, 6</td><td>80</td><td>C</td><td>4</td><td>116,594</td><td>2,066</td><td>125,210</td><td>4,56</td><td></td><td>10, 5</td></tr>
<tr><td>P6</td><td>1</td><td>171, 3</td><td>71</td><td>D</td><td>1</td><td>67,435</td><td>4,271</td><td>91,708</td><td>3,61</td><td></td><td>11, 1</td></tr>
<tr><td>P6</td><td>2</td><td>251, 5</td><td>80</td><td>C</td><td>4</td><td>111,203</td><td>5,085</td><td>127,364</td><td>3,99</td><td></td><td>12, 5</td></tr>
<tr><td>P6</td><td>3</td><td>118, 9</td><td>60, 5</td><td>E</td><td></td><td></td><td></td><td>53,407</td><td>3,05</td><td>2,26</td><td>10, 1</td></tr>
<tr><td>P6</td><td>4</td><td>120, 4</td><td>60</td><td>E</td><td></td><td></td><td></td><td></td><td></td><td></td><td></td></tr>
<tr><td>P6</td><td>5</td><td>134, 4</td><td>77, 5</td><td>C</td><td>4</td><td>98,870</td><td>4,78</td><td>121,268</td><td>4,01</td><td></td><td>11, 5</td></tr>
<tr><td>P6</td><td>6</td><td>160, 9</td><td>69, 5</td><td>D</td><td>1</td><td>82,351</td><td>3,764</td><td>68,517</td><td>3,65</td><td></td><td>11</td></tr>
<tr><td>P6</td><td>7</td><td>124, 5</td><td>61, 5</td><td>E</td><td></td><td></td><td></td><td>57,886</td><td>2,9</td><td>2,4</td><td>10, 3</td></tr>
<tr><td>P6</td><td>8</td><td>149, 9</td><td>78, 3</td><td>C</td><td>4</td><td>104,123</td><td>4,228</td><td>130,064</td><td>4,17</td><td></td><td>10, 8</td></tr>
<tr><td>P7</td><td>1</td><td>118, 5</td><td>61</td><td>E</td><td>2</td><td>68,155</td><td>2,446</td><td>55,135</td><td>3,38</td><td></td><td>10, 1</td></tr>
<tr><td>P7</td><td>2</td><td>117, 9</td><td>60, 5</td><td>E</td><td>1</td><td>56,834</td><td>4,546</td><td>52,374</td><td>3,31</td><td></td><td>8,2</td></tr>
<tr><td>P7</td><td>3</td><td>302, 4</td><td>85</td><td>B</td><td>3</td><td>173,314</td><td>3,066</td><td>114,258</td><td>3,81</td><td></td><td>11, 5</td></tr>
<tr><td>P7</td><td>4</td><td>155, 8</td><td>69</td><td>D</td><td>2</td><td></td><td></td><td>77,434</td><td>3,75</td><td></td><td>10, 6</td></tr>
<tr><td>P8</td><td>1</td><td>119, 7</td><td>65</td><td>D</td><td>6</td><td>73,439</td><td>1,788</td><td>32,567</td><td></td><td></td><td>14, 6</td></tr>
<tr><td>P8</td><td>2</td><td>140, 1</td><td>65</td><td>D</td><td>1</td><td>61,896</td><td>0</td><td>73,228</td><td>3,68</td><td></td><td>10, 2</td></tr>
<tr><td>P8</td><td>3</td><td>126, 5</td><td>62</td><td>D</td><td>2</td><td>48,583</td><td>1,52</td><td>70,146</td><td>3,68</td><td></td><td>8,9</td></tr>
<tr><td>P8</td><td>4</td><td>135,</td><td>63</td><td>D</td><td>1</td><td>56,289</td><td>2,308</td><td>70,52</td><td></td><td></td><td>10,</td></tr>
</table>

		0						3			0
PS	s	11s, 4	61, s	E	4	4s,6s1	г,3б1	sg,g6 0	3,sg		11, 0
PS	б	152, S	6S, s	D	3	sg,810	3,12S	82,43 s	3,84		11, s
PS	i	1A7, 3	66, s	D	4	б3,1бг	1,40g	i2,g4 S	3,i6		10, i
PS	S	16A, 0	68	D	1	i6,4S0	A,321	iS,i0 2	3,sg		10, i
PS	g	134, 1	6A, s	D				36,s3 6		3,g4	
PS	1 0	1s3, г	6S	D	г	i0,108	4,4ff	is,4s g	3,61		10, S
PS	1 1	1A8, s	бб	D	3			SS,3i 2	3,s6	0,i2	11, 1
PS	1 г	1sg, 3	68	D	3	66.3gs	A,12A	is,gs 3	3,i6		8,8
PS	1 3	1i2, 1	i0	D	3	i2,iS2	1,363	SS,SS 2	4,32		12, 2
PS	1 4	184, 4	i1	D	3			11g,i 82	4,28	0,42	11, 2
PS	1 s	1i6, 3	i0	D	г	6i,20i	s,436	gg,si s			10, S
PS	1 б	21A, 1	is	C	г			13A,8 3i	4,04	0,s	10, 4
Pg	1	1gi, i	i3, s	C	s	116,96 6	г,1бг	64,yes 4		0,s4	14, 1
Pg	г	S0,g	s4	E	3			46,s2 1	3,S3	0,6i	11, 6
Pg	3	13s, g	6s	D	1	61,S99	4,111	66,32 S	3,s6		10, g
Pg	4	11g. S	бг	D	1	43,33г	6,44g	ss,ss 2	3,63	0,ig	10, i
Pg	s	133, 0	бб	D	1	48.4si	S,349	6i,gs 1	3,i6	0,if	10, 2
Pg	б	104, s	ss	E	s	42,60S	2,38S	ss,3g s	3,84		11, 6
Pg	i	101, 3	sg	E	1			60,0g 6	4,02	0,s	11, 6
Pg	S	86	SA, s	E	г			s3,ig g	4,2s	0,3i	13, 6
Pg	g	110	63	D	s	64,60s	1,s44	36,36 S	4,43		13, 0
Pg	1 0	121, S	61	E	1	60,Sg0	4,i0g	s3.22 g	3,31		10, g
Pg	1 1	1s6, 0	6i	D	1	IS,26S	6,368	6S,i1 3	3,6S		12
Pg	1 г	1ss, s	6i	D	1	SS,642	2,i0s	S4,3i 3	3,ss	0,ii	11, S
Pg	1 3	16i, g	68	D	1	бб,г1г	s,63s	S4,s1 2	3,s1	1,04	10, 3
Pg	1 4	ig,S	ss	E	1			s1,44 3	3,sg	0,s2	g,s
Pg	1 s	136, s	6s, s	D	1	s1,s60	4,383	i1,12 4	3,i2	0,6g	10, 3
Pg	1 б	131, s	6s	D	3	61,sgi	s,261	ss,S6 6	3,g3		12, 0
Pg	1 i	1s2, i	66, s	D	3	66,3g6	s,268	ii,6i i			

P9	18	105, 5	61	E	3	51,453	2,674	47,908	3,65		9,8
P9	19	127, 2	62, 4	D	1			76,94	3,65	0,64	9,9
P9	20	159, 8	67	D	2	65,920	5,759	83,979	3,56		10, 7
P9	21	222	78	C	2	78,768	0,773	124,1 02		0,59	11, 1
P9	22	155, 9	71, 5	D	3	67,749	4,86	76,335	4,4	0,32	13, 7
P9	23	156, 7	71	D	1			93,378	3,74	0,79	10, 6
P9	24	95,5	58	E	2	45,822	3,257	42,654	3,51		11, 2
P9	25	124, 3	64	D	1			72,968	3,64	0,84	10, 7
P9	26	117	59, 5	E		55,031	1,905	52,481	2,94	2,6	10, 1
P9	27	103, 9	61, 5	E	3			71,371	3,86	0,54	11, 2
P9	28	84,9	56	E	2	31,675	2,255	47,455	3,65		11, 0
P10	1	257, 5	80, 5	C	1	119,886	5,939	123,776	3,95		9,9
P10	2	202, 1	76, 5	C	0	131,273	4,124	61,535	3,17		8,6
P10	3	190, 0	74	C	1	95,535	0,772	90,527	3,95		10, 9
P10	4	175, 7	71	D	2	78,975	2,113	86,656	3,83		10, 9
P10	5	170, 0	70	D	2	79,588	3,559	83,864	3,78		10, 8

Physicochemical analysis of Los Deseos valence orange - Part Three

PUNT 0	#	ESP 0g	e m m	CALIB RE	COLO R NTC	WEIGHT g CASCA RA	WEIGHT g SEMIL LA	WEIGHT g JUICE	PH JUG O	ACIDE Z JUICE %	°BR IX
P10	6	215, 1	75	C	2	95,645	3,735	108,348	3,9		11
P10	7	113, 9	61	E	1	57,245	3,58	47,588	3,7		9,2
P10	8	64,6 0	50	E							
P10	9	152, 6	67	D	3	68,356	2,99	76,803	3,73		12, 3
P10	10	215, 3	77	C	1			121,058	4,02	0,5	11, 1
P10	11	170, 0	71	D	0	98,289	3,856	64,349	3,12		9,4
P10	12	123, 6	64	D	1			64,948	3,87	0,67	10, 3
P10	13	172, 7	68	D				72,873	3,19	1,88	9,6
P10	14	187. 5	72	C	2			113,7 02	3,94	0,57	11, 9
P10	15	166, 3	70	D	1			107,35	3,87	0,6	10, 3
P10	1	365,	93	A	5	208,82	1,269	131,0	4,27		12,
	б	0				S		11			S
Rĺ2	i	g0,S	S6	E	г			дs,g3 g	3,19	0,5	12, б

Rí2	г	!3Д, S	бб	D	i	Sg,19i	Д,129	6S,2S 3	Д,0б		ii, г
Rí2	3	!0!	S6	E	i	6í,093	г,109		2,91		g,s
Rí2	Д	24i, g	S0	C	0	iSS,Sg S	S,0бД	13,дг	3,3i	0,5Д	g,3
P!г	S	132, Д	б3	D	г	SS,619	i^SS	1i,sg g	3,1i		i0, g
P!г	б	isg, г	б1, S	D	г	6S^0	3,iii	S2,21 s	3,6S		g,g
P!г	1	iSS, Д	бб, Д	D	i	93,31г	Д,05Д	53,19 g	3,21		s,g
P!г	S	129, г	б3	D	i			69.1í г	2,91	!,3б	g,3
P!г	g	г3г, 3	1S	C	3	119,90 S	5,393	i00,i 1Д			12, г
P!г	i 0	iSS, б	11	D					2,S1		i0, g
P!г	i i	119, г	10	D				^,1Д i	3,гг	гД	i0, S
P!г	i г	iS0, 3	б9	D	3	ss,ggд	3,gS3	S3,3S б	3,11		i0, 3
P!г	i 3	95,7	S6, S	E	Д	Дг,гбД	2,SSS	Д1.05 S	Д,59		!Д, б
P!г	i Д	sд,г	SS	E	г						
P!г	i S	SД,Д	SS	E	i				3,1б	0,бг	i0, б
P!г	i б	S1,2	S3, S	E	i			Д9.91 г	3,1Д	0,бД	i0, 1
P!г	i 1	sд,g	SS	E	г			S0,6S Д	3,6Д	0,Д5	ii, g
P!г	i S	i0S, 1	59	E	i			б3,Дб б	Д,00	0,бг	ii, 1
P!г	i g	S9,S	S6, S	E	г	дг,gos	Д,013	31,50 g	3,1		ii, 1
P!г	г 0	i0S, 3	S9, S	E	i			б0.19 S	3,1!	0,5г	i0, г
P!г	г i	í2S, 3	б3, S	D	i			11,ДД S	3,б5	0,бг	i0, г
P!г	г г	13б, S	бб, Д	D	i			Yes^ 3	3,1S	0,5г	ii, 0
P!г	г 3	1S6, 1	13, S	C	i	S6,11S	д,659	90,гг g	3,SS		g,3
P!г	г Д	гг0	1S, S	C	0	1ДД,гб г	3^S	61.6S g			
P!Д	i	2SS, s	Yes	C	г	í23.66 б	0,31б	1гД,5 2s	Д^1		i0, г
P!Д	г	г5Д, б	S2	C	г	1ДД,11 S	3,2SS	103,Д 13	3,1		i0
P!Д	3	309, б	S9	B	S	2í9,1S 3	3,1S2	ДS,S0 б	Д,б		д,б
P!Д	Д	2S3, Д	SS	B	г	i3S,iS Д	S^S	i30,i 6S			
P!Д	S	i9S, 0	1г	C	г	10,62S	0,0	112,5 S	5,3б		i0, Д
P!Д	б	г1Д, 6	S2	C	i				5,31	0,!г	i0, 6
P14	7	276, 1	85	B	3	113,67 3	4,848	122,7 16	4,69		10, 6
P14	8	154, 3	68, 5	D	0				3,39	1,26	10, 5

P14	9	143, 4	67	D	1	64,919	1,301	65,872	4,97		9,7
P14	10	222, 8	79	C	1	105,018	3,242	106,620	5,19		10, 3
P14	11	183, 4	72	C	1	90,024	0,431	91,627	3,73		9,8
P14	12	220, 7	77	C	2	76,080	4,899	126,567	5,37		10, 3
P14	13	270, 0	87	B	4	202,216	1,955	50,143	4,37		8,1
P14	14	275, 5	92	B	3	225,616	1,918	129,302	4,59		9,9
P14	15	257, 7	84	B	2	112,763	1,384	128,4 05	5,5		11, 1
P14	16	200	74	C	1	71,629	2,795	113,771	5,57		10, 5
P14	17	255, 8	84	B	1	120,93 0	2,406	113,211	3,81		10, 7
P14	18	178, 3	71	D	1				5,61	0,12	11, 1
P14	19	196, 3	73	C	3	89,845	1,986	96,226	4,97		10, 3
P15	1	187, 9	72, 5	C	4	86,625	2,436	94,783			9,8
P15	2	183, 4	70	D	3	80,516	7,434	89,366	3,75		10, 6
P15	3	157	68	D	3	65,997	2,727	79,633			9,6
P15	4	180, 7	70	D	4	65,450	2,401	108,679	3,67		10, 3
P15	5	164, 6	68	D	0	79,931	4,678	73,238			9,6
P15	6	188, 2	72	C	1	84,880	7,654	88,848	3,65		9,3
P15	7	193, 5	71	D	1	107,036	4,997	78,025	3,08		9,4
P15	8	152, 7	67	D							
P15	9	264, 2	72	C	3	84,559	6,073	106,694	3,69		10, 0
P15	10	245, 3	80	C	4	105,412	5,564	125,668	3,74		9,9

Physicochemical analysis of Los Deseos valence orange - Part 4

PUNT 0	#	PES Og	e m m	CALIB RE	COLO R NTC	WEIGHT g CASCA RA	WEIGHT g SEMIL LA	WEIGHT g JUICE	PH JUG O	ACIDE Z JUICE %	°BR IX
P15	11	229, 2	76	C	5	93,592	3,807	125,491	4,34		10, 8
P15	12	248, 7	79	C	3			128,404	4,24	0,44	9,9
P15	13	272, 8	81	C	2	119,748	5,68	123,793	4,26		10, 6

P1S	1 A	22g, g	?S	c	г	96,3A0	б,?04	11S,6 gS			10, g
P1S	1 S	214, б	A, S	c	г	sg,?2g	4,334	113,S 21			g,g
P1S	1 б	20s	S, S	c	1			111,0 A3	3,?S	1,0б	g,?
P1S	1 ?	113, 3	б0	E	1	S9,S06	2,63A	AS,sg A	2,g		g,s
P16	1	330	s, S	в	г	1?2,б4 ?	0,0s3	1A6.1 б	4,2?		10, g
P16	г	23s, 3	79	c	1	160.9s г	A,3SS	6S,22 A	3,24		s,1
P16	3	304, A	ss	в	г	1б3.13 б	4,42	12S,? 1б	A,2s		?,?
P16	A	2S1, г	s0, S	c	1	12s,S3 3	S,132	10g,3 0g	3,ss		11, б
P16	S	20s, A	?3, S	c	г	ss,?2S	1,1Sg	10S,g бб	4,02		g,g
P16	б	188, 3	?1, S	D	г	69Д61	3,г33	104,3 ?S	3,g2	0,S2	11, б
P16	?	1Ag, g	бб	D	г	sg,s01	3,s06	?g,1A 3	3,?s	0^g	14, 3
P16	s	1s1, ?	?г, A	c	A	?s,06S	1,Ass	92,1A 1			
P16	g	16?, 3	бg	D	г	6S,0sS	A,136	93.1s S	3,24	0,??	10, g
P16	1 0	16A, g	6s	D		ss,9S0	4,331	бб^ ?	2,?S	1,бб	g,A
P16	1 1	1S1, 0	б?	D	г	S3,AgS	1^0б	s9.36 g	3,S6	0,??	12, s
P16	1 г	gs,S	б1	E							10, 2
P16	1 3	gs,1	S?, S	E	г				3,g3	0,42	10, 0

ANNEX 2.

Physicochemical analysis of La Nueva Esperanza valencia orange

Physicochemical analysis of Valencia orange from La Nueva Esperanza - Part 1											
PUNT 0	#	PES Og	e m m	CALIB RE	COLO R NTC	WEIGHT g CASCA RA	WEIGHT g SEMIL LA	WEIGHT g JUICE	PH JUG O	ACID EZ JUICE % EZ	°BRI X
L	1	90,4	53	E	2	31,036	1,682	64,31 2	4,17	0,4	10,6
L	2	67,4	51	E	4	34,227	0,439	25,98 3	3,51		9
L	3	148, 4	65	D	3	58,754	1,92	95,29 8	3,97	0,52	10,3
L	4	135, 9	64	D	2	78,789	1,741	59,86 7	4,45	0,3	10,9
L	5	139, 0	64, 5	D	1	67,28	1,726	62,20 9	4,66		10,9
L	6	138, 6	64	D	1	60,435	1,194	67,03 6	4,62		9,5
L	7	128, 9	61	E	2	58,708	2,93	76,20 6	3,93	0,57	11,6
L	8	172, 0	70	D	1						
L	9	174, 6	70	D	2	80,687	2,638	97,09	3,6	0,82	10,1
L	1	138,	64	D	1	52,299	2,702	91,67	4,33	0,37	10,3
	0	g						A			
L	1 1	150, 0	61	D	1						
L	1 г	161, A	бg	D	3	64,606	4,344	s1,A6 3	3,sg		11,3
L	1 3	110, 1	6s	D	г	93,901	0,513	SA,SI 3	3,s1	0,62	9,6
L	1 A	166, 0	бб	D	1	67,7s	2,gsg	101,6 91			
L	1 S	163	бб	D	1	63,141	1,952	101,0 0g	4,17	0,42	9,3
L	1 6	11A, A	6s, S	D	б	11,siA	4,431	19,SA A	3,75		11,3
L	1 1	113, 3	бб, S	D	1				4,61	0,24	10,9
L	1 s	161, 0	69, 3	D	г	is,352	3,532	90,46 3	4,02	0,5	g,s
L	1 g	18G, 0	11, 1	D	г	yes,11A	3,HANDLE	96,05	3,64	0,69	9,3
L	г 0	1s0. 0	72, 3	C	г	69,102	6,9s9	106,4 1s	3,SA	0,59	9,9
L	г 1	1б3, 1	11, A	D	A	10,бОб	2,197	s0.9A 3	3,sg		g,s
L	г г	1s2	12	C	г	102,51 3	41,12	s0,60 1	4,03	0,44	10
L	г 3	1ff, 3	бg	D	1	s2,061	4,124	110,3 0s	3,ss	0,61	s,s
L	г A	1S6, 1	10	D	г	so,126	A,A2s	110,9	4,16	0,45	9,9
L	г S	11S	69, 3	D	1	61,si9	0,13S	112,2 13	3,12	0,14	9,5
L	г б	1s2	61	D	г	11,924	3,HANDLE	101,1 0A	3,s1	0,69	9,2

L	г 1	165, 3	61	D	3	65,sS9	4,121	77,9G 1	3,94		11,2
L	г s	1si, A	10	D	1	71,43	3,615	111.3 gg	3,if	0,54	g
L	г g	207, г	1S	C	г	102,13 б	3,405	10s,0 ss	3,sg	0,6	10,2
L	3 0	г3б	1S, A	C	1						
L	3 1	242, g	is, 3	C	1	136,92 S	1,559	114,6 S	3,ss	0,12	9,4
L	3 г	г1б, б	1S	C	г	113,11 1	4,13s	143.7 AS	4,31	0,42	9,4
L	3 3	гб0, 3	S6	B	1						
L	3 A	245, б	1S, A	C	1	92,211	4,309	150.0 AS	4,41	0,42	9,3
L	3 S	251, б	11, A	C	1	133,3б 3	2,101	124,9 31	3,14	0,11	9,4
L	3 б	264, 1	79, 1	C	1	106,39 g	3,625	156,1 2s	3,66	1,46	10,2
L	3 1	гго, S	13	C	г	100,13 1	4,314	125,0 s2	3,16		12,6
L	3 s	241, б	11	C	1	107,22 A	5,932		4,02		9,6
L	3 g	г30, 1	11	C	г	s0,AA1	2,559	ss,io s			
L	4 0	249, 2	79, 7	C	1	105,70 2	4,64	88,70 8	4,42		10,1
L	4 1	259, 2	77, 6	C	2						
L	4 2	295, 8	81, 6	C	1						
L	4 3	283, 7	81, 8	C	2	117,18 5	6,6	158,4 16	3,81	0,64	9,9
L	4 4	253, 4	78	C	1	125,64	4,012	130,1 39	3,87	0,62	9,6
L	4 5	265, 8	80, 3	C	3						
L	4 6	214, 6	85, 5	B	3						
L	4 7	212, 4	74, 4	C	2	76,009	0,91	139,4 75	3,84	0,62	10
L	4 8	224, 2	73, 7	C	2						
L	4 9	247, 7	76, 5	C	1	123,68 6	3,275	130,0 69	3,84	0,64	9,7
L	5 0	206, 7	71, 4	D	1	92,351	2,917	115,4 4		0,4	10,7
L	5 1	189, 1	69, 1	D	3	74,005	6,203		4,08		9,8
L	5 2		-							0,92	9,6
L	5 3	204, 6	75	C	3	100,69 7	3,861	102,2 56	3,94	0,67	10,7
L	5 4	296, 9	81, 5	C	1	127,25 7	4,497		3,56		12,5
L	5 5	259, 1	83	C	2	108,57 3	8,66	136,2 15	3,85		10,2
L	5 6	273, 4	82	C	1	136,74 6	4,96	128,9 36	3,88		10,7

L	57	261, 5	80	C	1	106,732	5,2	146,043			
L	58	217, 3	76	C	2	91,52	4,527	115,535	4,43		10,4
L	59	201, 9	73, 5	C	2	106,32	4,525	98,006	4,12	0,44	11,3
L	60	261, 7	81	C	2	115,11	4,617	135,62	3,84		10

Physicochemical analysis of La Nueva Esperanza Valencia orange - part 2

PUNT 0	#	ESP 0g	e m m	CALIB RE	COLO R NTC	WEIGHT g CASCA RA	WEIGHT g SEMIL LA	WEIGHT g JUICE	PH JUG O	ACID EZ JUICE %	°BRI X
L	61	256, 2	80	C	1	118,239	4,022	140,058	4,38	0,47	9,5
L	62	191, 3	71	D	1	91,21	3,874	101,613	3,92	0,6	10,3
L	63	283, 3	83	C	1		0,078				
L	64	255,9	82	C	6	128,906	5,538	116,57	3,69		11,8
L	65	297, 5	83, 5	C	1				4,38	0,4	9,3
L	66	238	76, S	C	2						
L	61	205, g	1A	c	3						
L	6S	1S1	11	c	1						
L	6g	1A0, 1	66	D	1	S1,AgS	1,616	S1,S16	A,11	0,A1	g,6
L	10	16S, A	бg	D	3				A,0S	0,A1	g,1
L	11	16A, S	бg	D	1	71,04	1,016	gg,61 6	A,SS	0,11	10,S
L	11	190, 1	1A, S	c	1	74,437	3,461	11S,A11	A,Sg	0,11	10,g
L	13	161, 1	6S, S	D	1	61,601	1,63S	100,S 1A	A,S1	0,31	g,1
L	1A	1g6, б	1A	c	1	S1,16A	3,414	114,6 g1	3,SS	0,11	10,6
L	1S	106, б	1S	c	1	101,63 б	1,gAg	10S,A Ag	A,1S	0.AA	g
L	16	101, 3	1S, S	c	1	g1,1g1	S,116	103.1 Ag	A,13		g,s
L	11	1A1, g	1S, S	c	1	9A,SSS	S,g61	1AS,111	A,11	0,A1	10,g
L	1S	1A1, 1	S0	c	1	6A,AS1	3,116	111.1 SA	3,gS	0,6A	10,1
L	1g	3A0, 1	SS	B	1						
L	S0	31A, б	SS	B	1						
L	S1	131, 0	16	c	1	HS,113	A,39A	143,10S	A,A1	0,3A	g,s
L	S1	116, 1	1A	c	1	g1,101	1,S3S	133.1 AS	A,3S	0,A1	g
L	S3	1g6, S	16	c	3	11,ASS	A,g1g	11g,S1	3,gS	0,AS	10,1
L	SA	1S1, g	бб	D	1				3,S3	0,6A	10,A

L	SS	1S3, g	61, S	D	1	61,113	1,g61	gS,11 A	3,1A	0,6g	g,1
L	S6	1SS, 1	11	D	1	gs,401	1,041	g1,1g6	A,1	0,S1	g,s
L	S1	1A1, S	61	D	1	6A,AS1	3,116	19.1s A	3,1S	0,6A	10,3
L	SS	111, 3	11	D	1	1S,611	1,114	g1,A1 g	3,AS	0,SA	S,S
L	Sg	10A, 1	1A, S	c	1	93,AS1	S,0A9	11A,1 1A	A,A1	0,31	10,1
L	g0	111, 0	16	c	1						
L	g1	103, 1	1S	c	1	SA,131	3,614	164,111	3,SS	0,61	10,S
L	g1	13S, A	1S	c	1	106,1	1,Sg	11S,0 6A	A,6A		g,g
L	g3	1A6, S	11, S	c	1	10S,S1 S	A,A9S	143.0 SS	3,SA	1,06	10,6
L	gA	11g	S1	c	1						
L	gS	1Sg, S	S0	c	1						
L	96	292, 1	83	C	1	142,272	3,653	141,596	3,63	0,84	11,8

ANNEX 3.

Analysis of Valencia orange peels (moisture and ashes)

ANALYSIS OF VALENCIAN ORANGE PEELS (MOISTURE AND ASHES)										
#	Element		X	Y	W2	Z Wet d		Z Ceniz a		
		W mues tra	W(chris ol)	W(c + m)		W3	W m. carbonized	W ceniz a	% HUMED AD	% ASH AS
3	87L	2,985	25,090	27,535	25,956	25,945	25,695	25,139	**65,031**	**2,004**
5	8P12	3,020	28,004	30,622		28,767	28,559	28,029	**70,856**	**0,955**
6	20P12	2,306	27,373	29,455		27,958	27,784	27,389	**71,902**	**0,768**
7	13P10	3,325	33,213	36,099		34,136	33,921	33,247	**68,018**	**1,178**
8	12P15	4,240	29,497	33,294		30,539	30,194	29,535	**72,557**	**1,001**
9	22P12	2,711	27,062	29,494		27,642	27,455	27,079	**76,151**	**0,699**
10	19P5	3,616	26,009	29,451	27,035	27,031	26,725	26,043	**70,308**	**0,988**
11	13L	3,564	27,305	30,634	28,126	28,122	27,908	27,34 0	**75,458**	**1,051**
12	11P5	3,565	30,386	33,814	31,398	31,398	31,152	30,427	**70,478**	**1,196**
13	11P12	4,585	29,900	34,313	31,08 4	31,079	30,822	29,949	**73,283**	**1,110**
14	3P6	4,345	27,839	32,005	29,059	29,055	28,776	27,894	**70,811**	**1,320**
15	20P5	5,548	27,479	32,679	29,14 0	29,139	28,572	27,539	**68,077**	**1,154**
16	7P9	2,621	30,386	32,844	31,31 0	31,311	31,028	30,421	**62,368**	**1,424**
17	11P8	4,821	27,373	31,932	28,663	28,663	28,224	27,416	**71,704**	**0,943**
18	21P12	3,298	27,839	30,912	28,764	28,764	28,481	27,862	**69,899**	**0,748**
19	8P5	3,377	30,274	33,513	31,564	31,565	31,216	30,326	**60,142**	**1,605**
20	15P10	4,141	33,213	37,162	34,481	34,481	34,095	33,257	**67,891**	**1,114**
21	4P12	3,021	24,887	27,733	25,90 0	25,900	25,517	24,914	**64,406**	**0,949**
22	17P5	6,585	29,900	35,975	32,254	31,878	31,849	29,975	**67,440**	**1,235**
23	18P14	3,757	28,004	31,425	29,196	29,186	29,185	28,047	**65,449**	**1,257**
24	10P10	6,696	29,497	35,644	31,702	31,510	31,499	29,575	**67,252**	**1,269**
25	17P14	7,153	27,062	33,606	29,487	29,036	28,993	27,147	**69,835**	**1,299**
26	4P5	8,928	27,305	35,621	31,01 7	30,186	29,905	27,431	**65,356**	**1,515**

27	23P9	4,715	26,009	30,335	27,69 0	27,670	27,658	26,07 7	**61,604**	**1,572**
28	16P15	5,168	25,090	29,756	26,627	26,590	26,573	25,151	**67,853**	**1,307**
29	17P12	4,236	20,601	24,384	22,00 1	21,968	21,962	20,636	**63,865**	**0,925**

Green: New Hope

White: Los Deseos

ANNEX 4.
Pectin yield from Los Deseos farm

<table>
<tr><th colspan="10">PECTIN YIELDS DESIRES</th></tr>
<tr><th>TRATA
LIE
0</th><th>r</th><th>W wet sample (g)</th><th>W dry sample (g)</th><th>W dried extracted pectin</th><th>Average dry pectin extracted</th><th colspan="2">%
PERFORMANCE
(W2/WI)*100</th><th colspan="2">YIELD
IN DRY BASE</th></tr>
<tr><td rowspan="3">TI</td><td>rl</td><td>250</td><td>66,575</td><td>6,5677</td><td rowspan="3">5,9703666
7</td><td rowspan="3">2,39 %</td><td>2,627</td><td rowspan="3">8,97 %</td><td>9,865</td></tr>
<tr><td>r2</td><td>250</td><td>66,575</td><td>4,8895</td><td>1,956</td><td>7,344</td></tr>
<tr><td>r3</td><td>250</td><td>66,575</td><td>6,4539</td><td>2,582</td><td>9,694</td></tr>
<tr><td rowspan="3">T2</td><td>rl</td><td>250</td><td>66,575</td><td>3,9362</td><td rowspan="3">3,7630333
3</td><td rowspan="3">1,51 %</td><td>1,574</td><td rowspan="3">5,65 %</td><td>5,912</td></tr>
<tr><td>r2</td><td>250</td><td>66,575</td><td>3,2104</td><td>1,284</td><td>4,822</td></tr>
<tr><td>r3</td><td>250</td><td>66,575</td><td>4,1425</td><td>1,657</td><td>6,222</td></tr>
<tr><td rowspan="3">T3</td><td>rl</td><td>250</td><td>66,575</td><td>5,576</td><td rowspan="3">4,7128</td><td rowspan="3">1,89 %</td><td>2,230</td><td rowspan="3">7,08 %</td><td>8,376</td></tr>
<tr><td>r2</td><td>250</td><td>66,575</td><td>4,6836</td><td>1,873</td><td>7,035</td></tr>
<tr><td>r3</td><td>250</td><td>66,575</td><td>3,8788</td><td>1,552</td><td>5,826</td></tr>
<tr><td rowspan="3">T4</td><td>rl</td><td>250</td><td>66,575</td><td>3,8573</td><td rowspan="3">3,7979333
3</td><td rowspan="3">1,52 %</td><td>1,543</td><td rowspan="3">5,70 %</td><td>5,794</td></tr>
<tr><td>r2</td><td>250</td><td>66,575</td><td>4,1967</td><td>1,679</td><td>6,304</td></tr>
<tr><td>r3</td><td>250</td><td>66,575</td><td>3,3398</td><td>1,336</td><td>5,017</td></tr>
<tr><td rowspan="3">T5</td><td>rl</td><td>250</td><td>66,575</td><td>5,8404</td><td rowspan="3">6,404433
33</td><td rowspan="3">2,56 2</td><td>2,336</td><td rowspan="3">9,62 0</td><td>8,773</td></tr>
<tr><td>r2</td><td>250</td><td>66,575</td><td>6,8293</td><td>2,732</td><td>10,258</td></tr>
<tr><td>r3</td><td>250</td><td>66,575</td><td>6,5436</td><td>2,617</td><td>9,829</td></tr>
<tr><td rowspan="3">T6</td><td>rl</td><td>250</td><td>66,575</td><td>2,408</td><td rowspan="3">2,7846</td><td rowspan="3">1,11 4</td><td>0,963</td><td rowspan="3">4,18
3</td><td>3,617</td></tr>
<tr><td>r2</td><td>250</td><td>66,575</td><td>3,0295</td><td>1,212</td><td>4,551</td></tr>
<tr><td>r3</td><td>250</td><td>66,575</td><td>2,9163</td><td>1,167</td><td>4,380</td></tr>
<tr><td rowspan="3">T7</td><td>rl</td><td>250</td><td>66,575</td><td>5,7894</td><td rowspan="3">5,179966
67</td><td rowspan="3">2,07 2</td><td>2,316</td><td rowspan="3">7,78 1</td><td>8,696</td></tr>
<tr><td>r2</td><td>250</td><td>66,575</td><td>5,1645</td><td>2,066</td><td>7,757</td></tr>
<tr><td>r3</td><td>250</td><td>66,575</td><td>4,586</td><td>1,834</td><td>6,888</td></tr>
<tr><td rowspan="3">T8</td><td>rl</td><td>250</td><td>66,575</td><td>3,2518</td><td rowspan="3">3,211066 67</td><td rowspan="3">1,28 4</td><td>1,301</td><td rowspan="3">4,82
3</td><td>4,884</td></tr>
<tr><td>r2</td><td>250</td><td>66,575</td><td>2,9456</td><td>1,178</td><td>4,424</td></tr>
<tr><td>r3</td><td>250</td><td>66,575</td><td>3,4358</td><td>1,374</td><td>5,161</td></tr>
</table>

ANNEX 5.

Pectin yield from the Nueva Esperanza farm

YIELD PECTINS NEW HOPE									
TRATA LIE O	r	W sample moisture (g)	W dry sample (g)	W dried extracted pectin	Average dry pectin extracted	YIELD = (W2/WI)*10 0		PERFORMANCE IN DRY BASE	
TI	rl	150	39,945	2,981			1,988		7,464
	r2	150	39,945	2,918	2,950	**1,967**	1,945	**7,385**	7,306
	r3	150	39,945				0,000		0,000
T2	rl	40	10,652	0,322			0,805		3,024
	r2	40	10,652	0,298	0,310	**0,775**	0,744	**2,909**	2,793
	r3	40	10,652	0,072*			0,179		0,673
T3	rl	40	10,652	1,186			2,964		11,131
	r2	40	10,652	1,050	1,062	**2,654**	2,626	**9,967**	9,861
	r3	40	10,652	0,949			2,373		8.911
T4	rl	40	10,652	0,403			1,008		3,785
	r2	40	10,652	0,433	0,418	**1,045**	1,083	**3,926**	4,066
	r3	40	10,652	0,288*			0,720		2,704
T5	rl	40	10,652	0,603			1,509		5,665
	r2	40	10,652	0,652	0,639	**1,598**	1,629	**6,001**	6,118
	r3	40	10,652	0,663			1,656		6,220
T6	rl	150	39,945	0,889			0,593		2,226
	r2	150	39,945	0,634	0,762	**0,508**	0,423	**1,907**	1,588
	r3	150	39,945				0,000		0,000
T7	rl	40	10,652	1,154			2,884		10,830
	r2	40	10,652	0,943	1,024	**2,560**	2,358	**9,614**	8,855
	r3	40	10,652	0,975			2,438		9,156
T8	rl	40	10,652	0,361			0,904		3,393
	r2	40	10,652	0,373	0,331	**0,828**	0,934	**3,109**	3,506
	r3	40	10,652	0,259			0,647		2,428

1values not evaluated

ANNEX 6.

Esterification degree of pectin from Nueva Esperanza peel.

DEGREE OF ESTERIFICATION FINCA NUEVA ESPERANZA												
TREATMENT TO	W. PECTIN(g)	N(NaO H)	VI my	V2 my	Kf	Ke	Kt	ED Esterification		methoxylated groups	ED AVERAGE	CORRECTION (-9)
T1R1	0,204	0,098	1,30	4,50	2,82 3	9,77 3	12,5 97	77,58 6	0,77 6	12,871		
T1R2	0,205	0,098	1,20	4,40	2,59 4	9,51 0	12,1 03	78,57 1	0,78 6	13,025	**78,479**	**69,479**
T1R3	0,204	0,098	1,15	4,40	2,49 8	9,55 6	12,0 54	79,27 9	0,79 3	13,136		
T2R1	0,206	0,098	1,30	4,60	2,79 6	9,89 3	12,6 89	77,96 6	0,78 0	12,931		
T2R2	0,207	0,098	1,20	4,40	2,56 8	9,41 8	11,9 86	78,57 1	0,78 6	13,025	**78,370**	**69,370**
T2R3	0,205	0,098	1,20	4,40	2,59 4	9,51 0	12,1 03	78,57 1	0,78 6	13,025		
T3R1	0,203	0,098	1,15	4,50	2,51 0	9,82 1	12,3 31	79,64 6	0,79 6	13,193		
T3R2	0,204	0,098	1,15	4,60	2,49 8	9,99 0	12,4 88	80,00 0	0,80 0	13,248	**80,227**	**71,227**
T3R3	0,205	0,098	1,10	4,70	2,37 7	10,1 58	12,5 35	81,03 4	0,81 0	13,409		
T4R1	0,207	0,098	1,25	5,20	2,67 5	11,1 30	13,8 05	80,62 0	0,80 6	13,344		
T4R2	0,208	0,098	1,20	4,90	2,55 6	10,4 37	12,9 93	80,32 8	0,80 3	13,299	**80,212**	**71,212**
T4R3	0,204	0,098	1,30	5,10	2,82 3	11,0 76	13,9 00	79,68 8	0,79 7	13,199		
T5R1	0,204	0,098	1,40	4,40	3,04 1	9,55 6	12,5 97	75,86 2	0,75 9	12,602	**76,657**	**67,657**
T5R2	0,206	0,098	1,30	4,50	2,79 6	9,67 8	12,4 74	77,58 6	0,77 6	12,871		

T5R3	0,204	0,098	1,35	4,40	2,93 2	9,55 6	12,4 88	76,52 2	0,76 5	12,705		
T6R1	0,205	0,098	1,30	4,40	2,81 0	9,51 0	12,3 19	77,19 3	0,77 2	12,810		
T6R2	0,203	0,098	1,20	5,20	2,61 9	11,3 49	13,9 68	81,25 0	0,81 3	13,442	**79,481**	**70,481**
T6R3	0,205	0,098	1,20	4,80	2,59 4	10,3 74	12,9 68	80,00 0	0,80 0	13,248		
T7R1	0,204	0,098	1,30	4,50	2,82 3	9,77 3	12,5 97	77,58 6	0,77 6	12,871		
T7R2	0,205	0,098	1,35	4,60	2,91 8	9,94 2	12,8 59	77,31 1	0,77 3	12,828	**77,494**	**68,494**
T7R3	0,204	0,098	1,30	4,50	2,82 3	9,77 3	12,5 97	77,58 6	0,77 6	12,871		
T8R1	0,206	0,098	1,20	4,30	2,58 1	9,24 8	11,8 29	78,18 2	0,78 2	12,964		
T8R2	0,204	0,098	1,20	4,50	2,60 6	9,77 3	12,3 80	78,94 7	0,78 9	13,084	**78,567**	**69,567**
T8R3	0,203	0,098	1,20	4,40	2,61 9	9,60 3	12,2 22	78,57 1	0,78 6	13,025		

ANNEX 7.

Degree of Esterification of the pectin from Los Deseos peels

DEGREE OF ESTERIFICATION OF DESIRES												
TREATMENT	W PECTINA	N(NaO H)	V1 mi	V2 mi	Kf	Ke	Kt	ED Sterification		methoxy groups two	ED PROMEDIO	CORRECTION (-9)
T1R1	0,2016	0,087	1,1	3,1	2,136	6,020	8,16	73,81	0,738	12,280	78,231	69,231
T1R2	0,2012	0,087	0,8	3,1	1,557	6,032	7,59	79,49	0,795	13,168		
T1R3	0,2088	0,087	0,8	3,5	1,500	6,563	8,06	81,39	0,814	13,465		
T2R1	0,2013	0,087	1,2	3,0	2,334	5,835	8,17	71,43	0,714	11,905	73,371	64,371
T2R2	0,2053	0,087	1,0	2,8	1,907	5,340	7,25	73,68	0,737	12,260		
T2R3	0,2002	0,087	1,0	3,0	1,956	5,867	7,82	75,00	0,750	12,466		
T3R1	0,2042	0,087	0,7	4,5	1,342	8,628	9,97	86,54	0,865	14,261	86,910	77,910
T3R2	0,2014	0,087	0,6	4,1	1,166	7,970	9,14	87,23	0,872	14,368		
T3R3	0,2085	0,087	0,6	4,0	1,127	7,511	8,64	86,96	0,870	14,325		
T4R1	0,2026	0,087	1,0	3,9	1,932	7,536	9,47	79,59	0,796	13,184	77,290	68,290
T4R2	0,2016	0,087	0,9	3,8	1,748	7,379	9,13	80,85	0,809	13,380		
T4R3	0,2083	0,087	1,2	3,0	2,255	5,639	7,89	71,43	0,714	11,905		
T8R1	0,2067	0,087	1,2	3,0	2,273	5,682	7,95	71,43	0,714	11,905	74,206	65,206
T8R2	0,2015	0,087	1,0	3,0	1,943	5,829	7,77	75,00	0,750	12,466		
T8R3	0,2064	0,087	1,0	3,2	1,897	6,070	7,97	76,19	0,762	12,653		
T6R1	0,2087	0,087	1,1	3,0	2,063	5,628	7,69	73,17	0,732	12,179	76,86	67,86
T6R2	0,2024	0,087	0,7	2,9	1,354	5,609	6,96	80,56	0,806	13,334		
T6R3	0,2055	0,087	1,2	0,8	2,286	1,524	3,81	40,00	0,400	6,828		
T5R1	0,2034	0,087	1,2	2,9	2,310	5,582	7,89	70,73	0,707	11,795	71,87	62,87
T5R2	0,2023	0,087	1,1	3,0	2,129	5,806	7,93	73,17	0,732	12,179		
T5R3	0,2027	0,087	1,5	3,8	2,897	7,339	10,24	71,70	0,717	11,947		
T7R1	0,2049	0,087	1,5	3,5	2,866	6,687	9,55	70,00	0,700	11,679	73,84	64,84
T7R2	0,2071	0,087	1,2	3,7	2,268	6,994	9,26	75,51	0,755	12,546		
T7R3	0,2029	0,087	1,2	3,8	2,315	7,332	9,65	76,00	0,760	12,623		

ANNEX 8.

Gelation time of extracted pectin

GELATION TIME OF THE EXTRACTED PECTIN	
REPEAT	TIME IN MINUTES
R1	14 MIN 45 SEC
R2	16 MIN 13 SEC
R3	18 MIN 32 SEC
AVERAGE	16 MIN 30 SEC

ANNEX 9.

Gelation potential test of pectin

GELIFICATION								
TREATMENT	REPEAT	**PH**	my solution	**BRIX**	gr Sugar	**%PECTINA**	gr Pectin	**DEGREE OF GELATION**
Tl	R1	2, 2	90	55	85,52	0,5	0,8820	96,960
	R2	2, 2	90	55	85,52	0,5	0,8820	96,960
T2	R1	2, 2	90	65	106,78	0,5	0,9888	107,985
	R2	2, 2	90	65	106,78	0,5	0,9888	107,985
T3	R1	2, 2	90	80	198,688	0,5	1,4507	136,961
	R2	2, 2	90	80	198,688	0,5	1,4507	136,961

T4	R1	2, 2	90	55	85,52	1,0	1,7729	48,237
	R2	2, 2	90	55	85,52	1,0	1,7729	48,237
T5	R1	2, 2	90	65	106,78	1,0	1,9877	53,721
	R2	2, 2	90	65	106,78	1,0	1,9877	53,721
		2						
T6	R1	2, 2	90	80	198,78	1,0	2,9170	68,146
	R2	2, 2	90	80	198,78	1,0	2,9170	68,146
T7	R1	2, 8	90	55	85,52	0,5	0,8820	96,960
	R2	2, 8	90	55	85,52	0,5	0,8820	96,960
T8	R1	2, 8	90	65	106,78	0,5	0,9888	107,985
	R2	2, 8	90	65	106,78	0,5	0,9888	107,985
T9	R1	2, 8	90	80	198,688	0,5	1,4507	136,961
	R2	2, 8	90	80	198,688	0,5	1,4507	136,961

T10	R1	2, 8	90	55	85,52	1,0	1,7729	48,237
	R2	2, 8	90	55	85,52	1,0	1,7729	48,237
T11	R1	2, 8	90	65	106,78	1,0	1,9877	53,721
	R2	2, 8	90	65	106,78	1,0	1,9877	53,721
T12	R1	2, 8	90	80	198,688	1,0	2,9160	68,136
	R2	2, 8	90	80	198,688	1,0	2,9160	68,136
		8						
T13	RI	3, 4	90	55	85,52	0,5	0,8820	96,960
	R2	3, 4	90	55	85,52	0,5	0,8820	96,960
T14	RI	3, 4	90	65	106,78	0,5	0,9888	107,985
	R2	3, 4	90	65	106,78	0,5	0,9888	107,985
T15	RI	3, 4	90	80	198,688	0,5	1,4507	136,961
	R2	3, 4	90	80	198,688	0,5	1,4507	136,961
T16	RI	3, 4	90	55	85,52	1,0	1,7729	48,237
	R2	3, 4	90	55	85,52	1,0	1,7729	48,237
T17	RI	3, 4	90	65	106,78	1,0	1,9877	53,721

	R2	3, 4	90	65	106,78	1,0	1,9877	53,721
T18	RI	3, 4	90	80	198,688	1,0	2,9160	68,136
	R2	3, 4	90	80	198,688	1,0	2,9160	68,136

ANNEX 10.

Testing the degree of pectin gelation

TEST OF THE DEGREE OF PECTIN GELATION							
REPEAT N	P H	my solution	ºBRI X	gr Sugar	% PECTI NA	gr Pectin	DEGREE OF GELATION
R1	3	100	65	106,788 1	0,0486	0,1005	1062,568
R2	3	100	65	106,773 5	0,0970	0,2007	532,005
R3	3	100	65	106,782 6	0,1451	0,3004	355,468
R4	3	100	65	106,795 2	0,1931	0,4001	266,921
R5	3	100	65	106,780 0	0,2415	0,5006	213,304
R6	3	100	65	106,779 5	0,2896	0,6005	177,818
R7	3	100	65	106,780 1	0,3376	0,7004	152,456
R8	3	100	65	106,781 5	0,3854	0,8001	133,460
R9	3	100	65	106,780 0	0,4335	0,9002	118,618
R10	3	100	65	106,781 0	0,4815	1,0005	106,728

ANNEX 11.

Pectin Titratable Acidity of Los Deseos pectin

Los Deseos pectin titratable acidity				
TREATMENT O	Normality NaOH	Volume of NaOH spent (ml)	sample weight (g)	FREE ACIDITY (meq. Free carboxyls/g) DESIRES
T1	0,087	0,9000	0,2039	0,3841

T2	0,087	1,0667	0,2023	0,4588
T3	0,087	0,6333	0,2047	0,2692
T4	0,087	1,0333	0,2042	0,4403
T5	0,087	1,2667	0,2028	0,5434
T6	0,087	1,0000	0,2055	0,4233
T7	0,087	1,3000	0,2050	0,5518
T8	0,087	1,0667	0,2049	0,4530

ANNEX 12.

Nueva Esperanza pectin titratable acidity

Pectin titratable acidity Nueva Esperanza				
TREATMENT	Normality NaOH	Volume of NaOH spent (ml)	sample weight (g)	FREE ACIDITY (meq. Free carboxyl/g) NUEVA ESPERANZA
T1	0,0985	1,2167	0,2043	0,5862
T2	0,0985	1,2333	0,2060	0,5895
T3	0,0985	1,1333	0,2040	0,5470
T4	0,0985	1,2500	0,2063	0,5965
T5	0,0985	1,3500	0,2047	0,6494
T6	0,0985	1,2333	0,2043	0,5943
T7	0,0985	1,3167	0,2043	0,6344
T8	0,0985	1,2000	0,2043	0,5782

ANNEX 13.

Equivalent weight pectin Los Deseos

Los Deseos pectin equivalent weight				
TREATMENT	Normal NaOH	Volume of NaOH spent (ml)	sample weight (g)	EQUIVALENT WEIGHT (mg/meq)
T1	0,087	0,9000	0,2039	2603,661
T2	0,087	1,0667	0,2023	2179,598
T3	0,087	0,6333	0,2047	3715,064
T4	0,087	1,0333	0,2042	2271,042
T5	0,087	1,2667	0,2028	2207,615
T6	0,087	1,0000	0,2055	2362,452
T7	0,087	1,3000	0,2050	1840,290
T8	0,087	1,0667	0,2049	1812,261

ANNEX 14.

Equivalent weight Nueva Esperanza pectin

Equivalent weight Nueva Esperanza pectin				
TREATMENT 0	Normal NaOH	Volume of NaOH spent (ml)	sample weight (g)	EQUIVALENT WEIGHT (mg/meq)
TI	0,0985	1,2167	0,2043	1705,772
T2	0,0985	1,2333	0,2060	1696,446
T3	0,0985	1,1333	0,2040	1828,209
T4	0,0985	1,2500	0,2063	1676,536
T5	0,0985	1,3500	0,2047	1539,809
T6	0,0985	1,2333	0,2043	1682,721
T7	0,0985	1,3167	0,2043	1576,220
T8	0,0985	1,2000	0,2043	1729,463

ANNEX 15.

Pectin yield at maturity stages 1, 2 and 3 of the Los deseos farm by the t4 method - 600w -10 minutes, dry samples and without essential oils.

PECTIN YIELD AT MATURITY STAGES 1, 2 AND 3 OF THE NUEVA ESPERANZA FARM BY METHOD T4-600W- 10 MINUTES, DRY SAMPLES AND WITHOUT ESSENTIAL OILS.									
	total weight m. + bag	Weight canti. For pectin	paper weight	weight for pei + pectin	pectin weight	initial pectin weight m.	% drypectin	Average % drypectin	standard deviation
the	85,60-1	10,12	1,08	1,42	0,34	2,84	6,11	**6,38**	**0,382**
Ib*	76,68-1	10,26	1,1	1,28	0,18	1,33	3,05		
1c	87,16-1	10,36	1,09	1,46	0,37	3,08	6,65		
2a	76,55-1	10,17	1,08	1,42	0,34	2,53	6,34	**6,30**	**0,057**
2B	74,74-1	10,2	1,1	1,46	0,36	2,6	6,26		
2C *	76,35-1	10,12	1,09	1,3	0,21	1,56	3,54		
3A	53,04-1	10,41	1,09	1,37	0,28	1,39	4,02	**4,07**	**0,071**
3B	36,32-1	10,08	1,08	1,44	0,36	1,26	4,12		
3C *	42,62-1	10,22	1,08	1,35	0,27	1,1	2,78		

*non-averaged values

ANNEX 16.

Pectin yield at maturity stages 1, 2 and 3 of the nueva esperanza farm by the t4 method - 600w -10 minutes, dry samples and without essential oils.

PECTIN YIELD AT MATURITY STAGES 1, 2 AND 3 OF THE NUEVA ESPERANZA FARM BY METHOD T4-600W- 10 MINUTES, DRY SAMPLES AND WITHOUT ESSENTIAL OILS.

	total weight m. + bag	weight canti. For pectin	paper weight	weight paper + pectin	pectin weight	initial pectin weight m.	% drypectin	average % drypectin	standard deviation
*	62,19-1	10,15	1,11	1,16	0,05	0,3	1,31	**7,56**	**0,000**
Ib	48,39-1	10,42	1,08	1,44	0,36	1,64	7,56		
1c *	71,77-1	10,08	1,13	1,17	0,04	0,28	1,12		
2a *	60,26-1	10,42	1,07	1,22	0,15	0,85	2,15	**7,65**	**1,485**
2B	51,33-1	10,53	1,05	1,55	0,5	2,15	6,6		
2C	75,26-1	10,29	1,08	1,55	0,47	3,43	8,7		
3A	32,67-1	10,02	1,1	1,44	0,34	1,11	4,66	**5,34**	**0,962**
3B	80,73-1	10,33	1,09	1,27	0,18	1,39	6,02		

*non-averaged values

ANNEX 17.

Moisture and ash determination of the pectin extracted from the two farms.

DETERMINATION OF MOISTURE AND ASHES OF PECTIN EXTRACTED FROM THE FARM LOS DESEOS (1) AND NUEVA ESPERANZAR)									
TRATAMIENTO	Lot	W Crucible	Wcrisol + sample	Initial sample W (g)	W2 crucible + dry sample	W3 crucible + dry sample	W crucible + Ashes	HUMIDITY (%)	ASH (%)
T1	1	20,926		0,685	21,581	21,580	20,950	4,526	3,670
	2	20,926	21,571	0,645		21,539	20,949	4,961	3,752
T2	1	15,469		0,563	16,005	16,005	15,488	4,796	3,545
	2	15,469	16,290	0,821		16,240	15,499	6,090	3,891
T3	1	13,584		0,716	14,267	14,266	13,603	4,749	2,786
	2	13,584	14,340	0,756		14,294	13,605	6,085	2,958
T4	1	20,590		0,615	21,178	21,177	20,612	4,553	3,748
	2	20,590	21,100	0,510		21,071	20,609	5,686	3,950
T5	1	27,297		0,601	27,869	27,868	27,318	4,992	3,678
	2	27,297	28,313	1,016		28,258	27,333	5,404	3,746
T6	1	23,592		0,632	24,192	24,191	23,613	5,222	3,506
	2	23,592	24,224	0,632		24,184	23,617	6,329	4,223
T7	1	24,885		0,821	25,664	25,663	24,912	5,238	3,470
	2	24,885	25,500	0,615		25,462	24,909	6,179	4,159
T8	1	27,371		0,845	28,178	28,177	27,402	4,615	3,846
	2	27,371	28,422	1,051		28,359	27,410	5,994	3,947

ANNEX 18.

Statistical analysis of pectin yields from the Los Deseos farm.

Pectin yield values wishes

Pectin yield LOS DESEOS								
	T1	T2	T3	T4	T5	T6	T7	T8
R1	9,865	5,912	8,376	5,794	8,773	3,617	8,696	4,884
R2	7,344	4,822	7,035	6,304	10,258	4,551	7,757	4,424
R3	9,694	6,222	5,826	5,017	9,829	4,380	6,888	5,161

Analysis of variance

Source	*Sum of Squares*	*Gl*	*Medium Square*	*F-Ratio*	*Falor-P*
Between groups	80,7672	7	11,5382	14,51	0,0000
Intra groups	12,7198	16	0,794989		
Total (Corr.)	93,487	23			

Multiple Range Testing

Method: 95.0 percent LSD

	Cases	*Media*	*Homogeneous Groups*
T6	3	4,18267	X
T8	3	4,823	X
T2	3	5,652	XX
T4	3	5,705	XX
T3	3	7,079	XX
T7	3	7,78033	XX
T1	3	8,96767	XX
T5	3	9,62	X

Contrast		*Difference*	*+/- Limits*
T1 - T2	*	3,31567	1,54331
T1 - T3	*	1,88867	1,54331
T1 - T4	*	3,26267	1,54331
T1 - T5		-0,652333	1,54331
T1 - T6	*	4,785	1,54331
T1 - T7		1,18733	1,54331
T1 - T8	*	4,14467	1,54331
T2-T3		-1,427	1,54331
T2-T4		-0,053	1,54331
T2-T5	*	-3,968	1,54331
T2-T6		1,46933	1,54331
T2-T7	*	-2,12833	1,54331
T2-T8		0,829	1,54331
T3 - T4		1,374	1,54331
T3 - T5	*	-2,541	1,54331
T3 - T6	*	2,89633	1,54331

T3 - T7		-0,701333	1,54331
T3 - T8	*	2,256	1,54331
T4-T5	*	-3,915	1,54331
T4-T6		1,52233	1,54331
T4-T7	*	-2,07533	1,54331
T4-T8		0,882	1,54331
T5 - T6	*	5,43733	1,54331
T5 - T7	*	1,83967	1,54331
T5 - T8	*	4,797	1,54331
T6-T7	*	-3,59767	1,54331
T6-T8		-0,640333	1,54331
IT7-T8	*	2,95733	11, 543311

* indicates a significant difference.

ANNEX 19.

Statistical analysis of Nueva Esperanza's pectin yields

Pectin yield values new hope

		Pectin yield NEW EN				PERANZA		
	T1	T2	T3	T4	T5	T6	T7	T8
R1	7,464	3,024	11,131	3,785	5,665	2,226	10,830	3,393
R2	7,306	2,793	9,861	4,066	6,118	1,588	8,855	3,506
R3			8,911		6,220		9,156	2,428

Analysis of variance

Source	*Sum of Squares*	*Gl*	*Medium Square*	*F-Ratio*	*Falor-P*
Between groups	176,4	7	25,2	51,22	0,0000
Intra groups	5,9045	12	0,492042		
Total (Corr.)	182,305	19			

Multiple Range Testing

Method: 95.0 percent LSD

	Cases	*Media*	*Homogeneous Groups*
T6	2	1,907	X
T2	2	2,9085	XX
T8	3	3,109	XX
T4	2	3,9255	X
T5	3	6,001	X
T1	2	7,385	X
T7	3	9,61367	X
T3	3	9,96767	X
Contrast	SiR	Difference	+/- Limits
T1 - T2	*	4,4765	1,52835
T1 - T3	*	-2,58267	1,39518
T1 - T4	*	3,4595	1,52835
T1 - T5		1,384	1,39518
T1 - T6	*	5,478	1,52835
T1 - T7	*	-2,22867	1,39518
T1 - T8	*	4,276	1,39518
T2-T3	*	-7,05917	1,39518
T2-T4		-1,017	1,52835
T2-T5	*	-3,0925	1,39518
T2-T6		1,0015	1,52835
T2-T7	*	-6,70517	1,39518
T2-T8		-0,2005	1,39518
T3 - T4	*	6,04217	1,39518
T3 - T5	*	3,96667	1,24789
T3 - T6	*	8,06067	1,39518
T3 - T7		0,354	1,24789
T3 - T8	*	6,85867	1,24789
T4-T5	*	-2,0755	1,39518
T4-T6	*	2,0185	1,52835
T4-T7	*	-5,68817	1,39518
T4-T8		0,8165	1,39518

T5 - T6	*	4,094	1,39518
T5 - T7	*	-3,61267	1,24789
T5 - T8	*	2,892	1,24789
T6-T7	*	-7,70667	1,39518
T6-T8		-1,202	1,39518
T7-T8	*	6,50467	1,24789

* indicates a significant difference.

ANNEX 20.

Statistical analysis for the degree of esterification of desires.

Values of the degree of esterification of the pectin of the desires

Esterification grade of pectin Los Deseos								
	Tl	T2	T3	T4	T5	T6	TJ	TS
RI	64,81 0	62,42 g	JJ,53 S	J0,sg 2	6l,J3 2	64,lJ 1	6l,00 0	62,42 g
R2	J0,4S J	64,6S 4	JS,23 4	7l,85 1	64,lJ 1	Jl,55 6	66,5l 0	66,00 0
R3	J2,3g 5	66,00 0	jj,gs J	62,42 g	62,6g S		6J,00 0	6J,lg 0

Analysis of variance

Source	*Sum of Squares*	*Gl*	*Medium Square*	*F-Ratio*	*Falor-P*
Between groups	466,957	7	66,7081	6,46	0,0012
Intra groups	154,998	15	10,3332		
Total (Corr.)	621,955	22			

Multiple Range Testing

Method: 95.0 percent LSD

	Cases	*Media*	*Homogeneous Groups*
T5	3	62,867	X
T2	3	64,371	XX
T7	3	64,8367	XX
T8	3	65,2063	XX
T6	2	67,8635	XX
T4	3	68,2907	XX
Tl	3	69,2307	X
T3	3	77,9097	X

Contrast	SiZ	*Difference*	*+/- Limits*
T1 - T2		4,85967	5,59433
T1 - T3	*	-8,679	5,59433
T1 - T4		0,94	5,59433
T1 - T5	*	6,36367	5,59433
T1 - T6		1,36717	6,25465
T1 - T7		4,394	5,59433
T1 - T8		4,02433	5,59433
T2-T3	*	-13,5387	5,59433
T2-T4		-3,91967	5,59433
T2-T5		1,504	5,59433
T2-T6		-3,4925	6,25465
T2-T7		-0,465667	5,59433
T2-T8		-0,835333	5,59433
T3 - T4	*	9,619	5,59433
T3 - T5	*	15,0427	5,59433
T3 - T6	*	10,0462	6,25465
T3 - T7	*	13,073	5,59433
T3 - T8	*	12,7033	5,59433
T4-T5		5,42367	5,59433
T4-T6		0,427167	6,25465
T4-T7		3,454	5,59433
T4-T8		3,08433	5,59433
T5 - T6		-4,9965	6,25465
T5 - T7		-1,96967	5,59433
T5 - T8		-2,33933	5,59433
T6-T7		3,02683	6,25465
T6-T8		2,65717	6,25465
T7-T8		-0,369667	5,59433

* indicates a significant difference.

ANNEX 21.

Statistical analysis for the degree of esterification Nueva Esperanza

Pectin esterification degree NUEVA ESPERANZA								
	T1	T2	T3	T4	T5	T6	T7	T8
R1	68,586	68,966	70,646	71,620	66,862	68,193	68,586	69,182
R2	69,571	69,571	71,000	71,328	68,586	72,250	68,311	69,947
R3	70,279	69,571	72,034	70,688	67,522	71,000	68,586	69,571

Analysis of variance

Source	*Sum of Squares*	*Gl*	*Medium Square*	*F-Ratio*	*Falor-P*
Between groups	33,0904	7	4,7272	5,53	0,0022
Intra groups	13,6748	16	0,854674		
Total (Corr.)	46,7651	23			

Multiple Range Testing

Method: 95.0 percent LSD

	Cases	*Media*	*Homogeneous Groups*
T5	3	67,6567	X
T7	3	68,4943	XX
T2	3	69,3693	XX
T1	3	69,4787	XX
T8	3	69,5667	XX
T6	3	70,481	XX
T4	3	71,212	X
T3	3	71,2267	X

Contrast	SiZ	*Difference*	*+/- Limits*
T1 - T2		0,109333	1,60019
T1 - T3	*	-1,748	1,60019
T1 - T4	*	-1,73333	1,60019
T1 - T5	*	1,822	1,60019
T1 - T6		-1,00233	1,60019
T1 - T7		0,984333	1,60019
T1 - T8		-0,088	1,60019
T2-T3	*	-1,85733	1,60019
T2-T4	*	-1,84267	1,60019
T2-T5	*	1,71267	1,60019
T2-T6		-1,11167	1,60019
T2-T7		0,875	1,60019
T2-T8		-0,197333	1,60019
T3 - T4		0,0146667	1,60019
T3 - T5	*	3,57	1,60019
T3 - T6		0,745667	1,60019
T3 - T7	*	2,73233	1,60019
T3 - T8	*	1,66	1,60019
T4-T5	*	3,55533	1,60019
T4-T6		0,731	1,60019
T4-T7	*	2,71767	1,60019
T4-T8	*	1,64533	1,60019
T5 - T6	*	-2,82433	1,60019
T5 - T7		-0,837667	1,60019

T5 - T8	*	-1,91	1,60019
T6-T7	*	1,98667	1,60019
T6-T8		0,914333	1,60019
T7-T8		-1,07233	1,60019

* indicates a significant difference.

ANNEX 22.

Statistical analysis for the degree of methoxylation of desires.

Values of the degree of esterification of the pectin of the desires

Degree of pectin methoxylation Desires								
	T1	T2	T3	T4	TS	T6	T7	TS
RI	12,28 0	11,90 S	14,26 1	13,18 4	11,79 S	12,17 g	11,67 g	11,90 S
R2	13,16 S	12,26 0	14,36 S	13,38 0	12,17 g	13,33 4	12,54 6	12,46 6
R3	13,46 S	12,46 6	14,32 S	11,90 S	11,94 7		12,62 3	12,6S 3

Analysis of variance

Source	*Sum of Squares*	*Gl*	*Medium Square*	*F-Ratio*	*Falor-P*
Between groups	11,335	7	1,61928	6,38	0,0013
Intra groups	3,80494	15	0,253662		
Total (Corr.)	15,1399	22			

Multiple Range Testing

Method: 95.0 percent LSD

	Cases	*Media*	*Homogeneous Groups*
T5	3	11,9737	X
T2	3	12,2103	XX
T7	3	12,2827	XX
T8	3	12,3413	XX
T6	2	12,7565	XX
T4	3	12,823	XX
T1	3	12,971	X
T3	3	14,318	X

Contrast	SiZ	Difference	+/- Limits
Tl - T2		0,760667	0,876513
IT - T3	*	-1,347	0,876513
Tl - T4		0,148	0,876513
Tl - T5	*	0,997333	0,876513
Tl - T6		0,2145	0,979971
Tl - T7		0,688333	0,876513
Tl - T8		0,629667	0,876513
T2-T3	*	-2,10767	0,876513
T2-T4		-0,612667	0,876513
T2-T5		0,236667	0,876513
T2-T6		-0,546167	0,979971
T2-T7		-0,0723333	0,876513
T2-T8		-0.131	0,876513
T3 - T4	*	1,495	0,876513
T3 - T5	*	2,34433	0,876513
T3 - T6	*	1,5615	0,979971
T3 - T7	*	2,03533	0,876513
T3 - T8	*	1,97667	0,876513
T4- T5		0,849333	0,876513
T4- T6		0,0665	0,979971
T4- T7		0,540333	0,876513
T4- T8		0,481667	0,876513
T5 - T6		-0,782833	0,979971
T5 - T7		-0,309	0,876513
T5 - T8		-0,367667	0,876513
T6- T7		0,473833	0,979971
T6- T8		0.415167	0,979971
T7- T8		-0,0586667	0,876513

* indicates a significant difference.

Values of the degree of esterification of Nueva Esperanza pectin.

ANNEX 23.

Statistical analysis for the degree of methoxylation of Nueva Esperanza.

Degree of pectin methoxylation Nueva Esperanza								
	Ti	T2	T3	T4	TS	T6	T7	TS
R1	12,S7 i	12,g3 1	13,19 3	13,34 4	12,60 2	12,81 0	12,S7 1	12,g6 4
R2	13,02 S	13,02 S	13,24 S	13.2g g	12,S7 1	13,44 2	12,S2 S	13,0S 4
R3	13,13 6	13,02 S	13,40 g	13,19 g	12,70 S	13,24 S	12,S7 1	13,02 S

Analysis of variance

Source	*Sum of Squares*	*Gl*	*Medium Square*	*F-Ratio*	*Falor-P*
Between groups	0,805726	7	0,115104	5,54	0,0022
Intra groups	0,332439	16	0,0207774		
Total (Corr.)	1,13816	23			

Multiple Range Testing

Method: 95.0 percent LSD

	Cases	*Media*	*Homogeneous Groups*
T5	3	12,726	X
T7	3	12,8567	XX
T2	3	12,9937	XX
TI	3	13,0107	XX
T8	3	13,0243	XX
T6	3	13,1667	XX
T4	3	13,2807	X
T3	3	13,2833	X

Contrast	*Sig.*	*Difference*	*+/- Limits*
T1 - T2		0,017	0,249498
T1 - T3	*	-0,272667	0,249498
T1 - T4	*	-0,27	0,249498
T1 - T5	*	0,284667	0,249498
T1 - T6		-0,156	0,249498
T1 - T7		0,154	0,249498
T1 - T8		-0,0136667	0,249498
T2-T3	*	-0,289667	0,249498
T2-T4	*	-0,287	0,249498
T2-T5	*	0,267667	0,249498
T2-T6		-0,173	0,249498
T2-T7		0,137	0,249498
T2-T8		-0,0306667	0,249498
T3 - T4		0,00266667	0,249498
T3 - T5	*	0,557333	0,249498
T3 - T6		0,116667	0,249498
T3 - T7	*	0,426667	0,249498
T3 - T8	*	0,259	0,249498
T4-T5	*	0,554667	0,249498
T4-T6		0,114	0,249498
T4-T7	*	0,424	0,249498
T4-T8	*	0,256333	0,249498
T5 - T6	*	-0,440667	0,249498
T5 - T7		-0,130667	0,249498
T5 - T8	*	-0,298333	0,249498
T6-T7	*	0,31	0,249498
T6-T8		0,142333	0,249498
T7-T8		-0,167667	0,249498

* indicates a significant difference.

Printed by Books on Demand GmbH, Norderstedt / Germany